CREATION UNFOLDING

A NEW PERSPECTIVE ON *EX NIHILO*

K.P. COULSON

Creation Unfolding: A New Perspective on Ex Nihilo

Published in 2020 by Phaneros Press

ISBN 978-1-7345842-0-2 (print)
ISBN 978-1-7345842-1-9 (ebook)

Cover design by Kristen Ingebretson
Interior design by Michael Grossman

I dedicate this book to my loving wife for the investment of her precious time, energy, and patience.

And to my three little pumpkins who lovingly embezzled all three of these resources.

Table of Contents

Acknowledgments

First, I want to thank my Lord and Savior, Jesus Christ. Without Christ, we can do nothing (John 15:5). This book is the result of much prayer and a clear objective to glorify God and expand His Kingdom. Nothing else in life matters without an eternal perspective.

Next, I want to thank my lovely wife and best friend, Beth. She has provided an insightful mind, a helpful ear, a guiding eye and a patient heart throughout the entire process. Truly, this book could not have been written without her. I want to thank my good friend Dr. Matt McLain for providing many insightful ideas along the way and for spurring me on to put pen to paper. Thank you to Donna Messenger for assistance with editorial work. Thank you to my long-time pastor, Dr. Craig Lloyd; my physics professor, Dr. Steve Gollmer; my geology professor, Dr. John Whitmore, and my friend and colleague, Dr. Sarah Maithel for assistance with manuscript revision.

I also want to thank numerous friends and colleagues who contributed with a word here or an idea there. As you read through the book, no doubt you will see your signature in a phrase, a reference or a concept.

Finally, I want to thank San Diego Christian College for giving me the much-needed time to write and do research.

Introduction

Most Christians have little problem explaining how things work in the natural realm. Articulating processes related to tidal cycles, lunar phases or solar eclipses are rudimentary fare from a very young age. Things get a little trickier, however, as we move back in time to the solar system's inception.

How should Christian laypeople and educators, for example, *scientifically* understand the processes at work during the *supernatural* formation of the Earth, the moon, the sun and, in fact, the entire universe? Is this even possible?

Questions such as these become even more complex for Christians who are also scientists. How, for example, should the Christian geophysicist approach his discipline when he believes that the Earth's fundamental internal systems were created supernaturally? This is especially frustrating when his secular peers make successful predictions about the Earth's interior, even though those predictions are based on data retrieved from processes set in motion during the Earth's supernatural construction.

This book presents a scientific and theological synthesis that both affirms a six calendar-day creation while at the same time providing a solid scientific methodology[1] from which the Christian layperson, educator, and scientist can approach the origin of the universe.

These ideas will enable the scientist to do good, predictive science while remaining faithful to a literal interpretation of the biblical text. They will enable the educator to authentically and passionately portray science in a positive light while rejoicing in a powerful, all-wise Creator. Finally, they will ease some of the tension that currently exists between science and faith, allowing for God's purpose in creation to take its full effect on the hearts and minds of family members, friends, students, neighbors, and work colleagues.

As the synthesis in this book unfolds, it will also address many common misconceptions in modern young-Earth creationism. These misconceptions are based on a creationist apologetic that fails to distinguish the miracle of creation from the natural laws that uphold that creation. The universe was made in literally six days, a statement that can be accepted only with the eyes of faith (Hebrews 11:3). This is because humans know by experience that planets, stars, and solar systems do not form in six days using *known* rates of natural change. The creation of the universe was clearly a miracle and, as such, must be treated as one by those who hold to the authority of God's revelation in Scripture.

Equally relevant, however, are the natural laws that currently sustain the creation. This juxtaposition between miracle on the one hand and natural process on the other, if not clearly understood, can lead to an emphasis in one direction over the other that ends up weakening either theology or science. Consider the well-meaning

---

1 I am presupposing that some kind of objective, scientific method does indeed exist. This presupposition is not as robust as some might suggest. See Moreland (1989) for a good discussion on the merits of something we call, *the* scientific method.

Christian who falls into an origins-related "evidence-based" trap. By "evidence-based," I am referring to the common misconception among evangelical Christians that assumes God's supernatural acts during Creation Week are somehow *scientifically detectible*, and that it is only the non-Christian's scientific bias that leads them to believe evolutionary and/or old-Earth ideas.

As we shall see, there are many aspects of creation that, without deferring to special revelation, *will only lead* to an evolutionary and/or old-Earth perspective. Quite simply, this is because God created a mature, functioning universe. As such, the Christian should not push science (a dependence on natural process) back into Creation Week *for the purpose of proving* that the Earth is young. Theologically speaking, the Scriptures nowhere prescribe a gospel apologetic that involves "proving" that the universe is 6,000 years old. Although it is true that the heavens "declare the glory of God," nowhere in Scripture does such a declaration move people into a personal relationship with God (Romans 1). Only the gospel can do that. In contrast, the Scriptures seem to imply that God has built a measure of ambiguity into the creation that leaves room for unbelief.

From a scientific perspective, God created the universe in *anticipation* of the very laws that would be required to sustain it. This would seem to suggest that the processes we see at work within the universe today were also used to create the universe in the first place. The natural rhythm of these processes is, of course, essential for the believing scientist as he seeks to obey God's mandate to have dominion and rule over that which God has made (Genesis 1:26–28).

How can Christians who are scientists, or anyone for that matter, fulfill this mandate if God used completely different processes to create the universe than He currently uses to sustain it? The obvious implication from this statement is that God did not. This means that Christians who are also scientists should be able to apply scientific methodology to origins-related processes *for the purpose of* doing good, predictive science, while at the same time remaining faithful to a literal interpretation of Genesis—a position that can be affirmed only by faith. What all Christians must guard against is an overemphasis on *origins-related* processes for the *sole* purpose of *proving* that the Earth is young.

The thesis presented in this book is the product of many years of deep, critical reflection about science and the way science interfaces with young-Earth creationism. There are many reasons for this, but it primarily results from my own thoughtful and skeptical personality. I think of the words of Socrates who said, "the unexamined life is not worth living." Those who know me know that I examine *everything* and do so with analytical rigor, whether that be in areas related to theology or science. Before embarking on a scientific career, I had the great privilege of studying theology at a conservative Bible college and equally conservative seminary,[2] both of which fostered a mindset of critical thinking. This valuable experience has enabled me to grasp the theological facets of creationism and tie them nicely with the scientific facets, themselves garnered from a further eight years of study in geology that culminated in an earned PhD in earth science. Experience drawn from this diverse academic background, in conjunction with the publication of numerous research-based articles in both the secular and creationist scientific communities, has allowed me to write what I believe to be a powerful apologetic in defense of *biblical* creationism.

---

2 I did not graduate from seminary. As a new believer in Christ, I was thankfully encouraged by my pastor to spend a year at seminary for the purpose of acquiring a theological foundation upon which to set my burgeoning faith.

Chapter 1

"Houston, We Have a Problem"

The problem

As a scientist who is also a Christian, I have had a solid vantage point that has enabled me to see the relevance of the creation/evolution debate. I am unreservedly and unashamedly a six-day creationist, in that I firmly believe God created the entire cosmos in literally six 24-hour days.

As a scientist, an educator and a lay minister, however, I have become increasingly aware of some glaring problems within mainstream young-Earth creationism that stem from what I believe is an overly suspicious approach to conventional science. Exacerbating the problems are a plethora of online young-Earth creationist (YEC)[1] websites that are hosted by well-meaning Christians who are neither scientists nor theologians. These websites claim that scientific evidence abounds for the youth of the creation, and that a belief in millions of years is nothing more than propaganda cooked

1 Young-Earth creationists believe that the Earth was made in six, literal days and that Adam and Eve were literal people who experienced a literal Fall into sin.

up by biased evolutionists out to wantonly deceive the world at large. In their minds, a YEC interpretation of nature is *obvious*, and only a fool would fall for the secular view that believes the Earth is 4.6 billion years old.

Unfortunately, these ideas have trickled into the church at large, and as a result, incite evangelical Christians to say things like, "there is not a shred of evidence in support of an evolving universe," or "how could anyone believe in millions of years of Earth history?" It may surprise many Christians to learn, however, that much evidence exists in nature that supports the antiquity of the Earth, the solar system and the universe. I believe that many Christians growing up in Christian homes leave the faith because they have not been adequately prepared to deal with this reality, and unless Christian parents and educators know how to handle conflicting evidence *biblically*, they will not effectively prepare young people for the onslaught awaiting them in many areas of higher education.

Working towards a solution

In proposing a solution, we must first acknowledge that evidence *does* exist supporting the antiquity of the Earth. I will flesh this out in later chapters, but for now, consider the geophysical processes that are presently working to sustain our planet. The continental crust, for example, is made of a lighter rock than that found in the mantle (the Earth is broken into three basic parts: the crust, the mantle, and the core). Geophysicists believe that lighter, continental crustal rocks have "evolved" from heavier, basaltic oceanic crustal rocks,[2] a notion supported by present processes whereby lighter, continental rock

2 Both types of crustal rocks are associated with the crust. One type covers the ocean floor, the other sits beneath our feet.

grows at a rate of about 1 cubic kilometer per year (Baumgardner 2000, p. 58).

By estimating the current volume of the Earth's continental rocks, and then extrapolating back the *current* rate at which these rocks are being formed, one can easily approximate a time frame for the Earth's formation in the millions and/or billions of years.

This is not an isolated case. Within the Earth's core and mantle there presently exist multiple geological structures that, *given today's rates of change*, can only have come into existence over very long time frames. Simplistically, planets like Earth do not come into existence in six normal days. The presence of vast processes in conjunction with short time frames can, therefore, only be understood by appealing to some kind of *mature creation* apologetic. This apologetic, however, has a number of obvious drawbacks especially as it interfaces with scientific methodology. Many scientists, as one might expect, tend to think in very rational terms. It is not easy to see a group of relationships that exist within a Creation Week rock, for example, and feel as though you really cannot analyze those relationships. What's the point since the rock was made supernaturally and the relationships that exist within the rock are not real?

There is also the problem of intrinsic properties. Should supernaturally created objects *look* like their naturally made equivalents, or do they have *special* constituents that identify their having a supernatural origin? These problems have led non-Christians, and even Christians, to suggest that a commitment to mature creation apologetics is a science stopper. If Christians accept these ideas, then how can Christians who are also scientists be expected to do methodical, predictive science in areas such as astronomy, geophysics and astrophysics, since the objects they are studying had a supernatural origin?

Similarly, appealing to the supernatural will eventually lead to "gaps" in our understanding that can only be filled with God. In stark contrast, some YEC believe that a commitment to mature creation apologetics will weaken scientific research validating the *youth* of the universe! In their minds, we need to use science to prove that a literal interpretation of Genesis is the right one. Perhaps one of the major objections to a mature creation apologetic is the concept of *deception* on God's part. This accusation comes from old-Earth creationists (OEC),[3] theistic evolutionists (TE),[4] and *even* many YEC. These issues are a fair appraisal of mature creation apologetics. My hope is to dispel these objections and counter each of them with a biblically sound and scientifically solid argument that ties mature creation apologetics, the Bible, and science together into a coherent synthesis.

This synthesis will not only alleviate some of the tension that exists between science and mainstream young-Earth creationism, it will also provide a platform from which Christian astronomers, astrophysicists, and geophysicists can interface with the *secular* scientific community. This is important because not all Christian scientists are doctors, chemists, engineers, and environmental specialists. Many are also doing *secular* research in areas of geophysics and cosmology—disciplines that are profoundly related to issues of origins. Of course, a commitment to a mature creation apologetic sounds like a scientific contradiction in terms—how can you do good, predictive science on

3 Old-Earth creationists believe that the Earth is billions of years old, but generally do not believe in Darwinian evolution. They believe in a literal Adam and Eve and a literal Fall into sin.

4 Theistic evolutionists completely accept the secular view of Earth history and Darwinian evolution. They do not believe in a literal Adam (made from the dust) and Eve (made from Adam), nor do they believe in a literal Fall into sin.

rocks that were assembled *supernaturally*? Traditional solutions have failed to solve this issue. This book, however, furnishes Christians with a fresh philosophical aspect that does, indeed, allow Christians to study supernaturally created rocks at the *secular* level.

God's design in Scripture

This book is not about "proving" young-Earth creationism right. In order to do this, I would be fighting against God's design in Scripture. That is right: God's design is not to prove that the Earth or the universe is 6,000 to 10,000 years old.

In 1 Corinthians 1:21 (ESV), Paul makes God's design abundantly clear: "For since, in the wisdom of God, the world did not know God through wisdom, it pleased God through the folly of what we preach to save those who believe." The world is not allowed to come to know the God of creation by pursuing its own wisdom (e.g., through scientific endeavor). This is by God's sovereign design. Think about it; if God really wanted to prove, *without a shadow of a doubt*, that He was the Creator, then He could have stamped, "Made by the God of the Hebrews" onto the buttocks of every child ever born using the language of every nation that ever existed. This would be *unambiguous*. It turns out that there are many theological reasons why God has chosen to veil His authoritative title as Creator. Not to *hide* it, but to *veil* it. Things are ambiguous for a reason (unpacked in Chapter 8).

Obviously, this kind of thinking is contrary to the approach many Christians take when trying to "defend" God's creative attributes. I sincerely understand this approach, but the zeal with which we focus on creation-related apologetics must be tempered by God's design in Scripture. This zeal goes both ways. Many YEC make the mistake of supposing that the universe is not only young, and that science proves this, but that God's purpose in Scripture profoundly agrees with this.

On the flipside, many OEC and TE do *exactly* the same thing; only in their minds, the universe is old, science proves this, and God's purpose in Scripture profoundly agrees with their position instead.

Both positions, however, are not handling the Word of God carefully. This book aims to flesh out God's *revealed* purpose in general revelation and rightly relate that purpose to God's design in special revelation. A correct perspective on these issues is crucial for the Christian whose number one goal is not to win arguments but to win converts to Christ.

What this book is *not* about

Many YEC publications are strictly apologetical in terms of the *science*. On display in these books are many scientific evidences that powerfully support the notion of a young Earth. That is not the purpose of this book. This book is a philosophical and theological treatise that seeks to dovetail the latter two disciplines with good scientific methodology.

Although I will use some scientific evidences in support of a young Earth, I will do so only for the purpose of supporting contextual arguments. I will also use a literal approach to the text of Scripture, but I will not seek to defend the text against a more liberal and/or analogical interpretation unless the context demands it.[5] Many books have already been written that address both the scientific credibility of a young Earth and the theological trustworthiness

5 Particularly related to Genesis 1 – 11.

of a faithful understanding of Genesis.[6] This book is entirely different in that it provides a workable framework for young-Earth creationism in general.

Caveats

Since I am a physical scientist, I will attempt to address only YEC issues related to the inorganic universe, such as the origin and age of the Earth, the solar system and the universe. Only vaguely will I touch on the life sciences and issues related to Darwinian evolution. I do believe that many of the ideas presented in this book are directly applicable to biological origins issues, but I have not attempted to flesh out those ideas here.

Having said all of this, I am fully aware that some of my scientific and/or philosophical ideas will not ally themselves with mainstream young-Earth creationism. These ideas are not a departure from the latter, but they certainly do seem to stretch the bounds of what is and what is not considered "orthodox." My deviation from mainstream young-Earth creationism is, however, related only to science and philosophy, *not* theology. I am and always will be committed to the full, plenary inspiration of the holy Scriptures, as well as the grammatical/historical method of biblical interpretation.

Throughout the book, I will use the term "naturalistic" to emphasize God's providential rhythms of regularity in nature. God uses *supernatural*, non-rhythmic creative powers. But, according to Scripture, He also uses natural, rhythmic sustaining powers. This is what I mean by naturalistic. Do not confuse naturalistic with

---

6 Brand and Chadwick (2016) is a particularly good resource. You can get the free eBook edition at this link: https://isgenesishistory.com/faith-reason-earth-history/

naturalism. The latter term denies supernatural events and divine superintendence in nature. When I use the term "naturalistic," I am in no way endorsing a materialistic, evolutionary world view.

Finally, don't let the science stop you reading the entire book. Although I have tried to keep things simple, some of the scientific concepts do tend to get a little deep in the latter parts of chapters 5 and 6. Keep going! Skim and come back if you must, but don't give up! The pace drops in Chapter 7, slowing to a leisurely stroll for the rest of the book as the science gives way to theology.

Chapter 2

Mature Creation Apologetics

Prior to the 19th century

Early Christian thinkers such as Augustine of Hippo, Basil of Caesarea, and Thomas Aquinas, among many others, clearly articulated a young-age view of the creation (Millam 2011; Sibley 2012) such that their writings did not require a mature creation apologetic. It is true, many of these early thinkers differed as to the *duration* and/or the *literalness* of the days in Genesis 1,[1] but as OEC John Millam admits, all of them, without exception, believed the Earth to be less than about 10,000 years old (Millam 2011).

The 19th century

A cogent, well-grounded and persuasive mature creation apologetic did not appear until 1822, when Granville Penn, a well-known

---

1 Most held to a calendar-day view for the days, but a few, including Augustine, seemed to believe that the entire cosmos was created instantly.

and respected scriptural geologist,[2] wrote *A Comparative Estimate of the Mineral and Mosaical Geologies*. Penn, like many of his contemporaries, was trying to come to terms with the growing body of geological evidence that seemed to support the notion of an ancient Earth.

An area receiving a great deal of attention at the time was related to the nature and origin of granitic continental rocks (called "primary" rocks). It was obvious to early investigators that these large igneous bodies were composed of crystals that must have precipitated from some kind of primordial liquid, much like salt crystals precipitate from brine (Penn 1825, p. 42). Those who subscribed to these views took the name *mineral geologists*, the word "mineral" underscoring their adherence to a precipitated origin for granitic crystals.

Not only did the mineral geologists put forward the idea of a primordial fluid from which these crystals precipitated, they went so far as to suggest that the processes responsible for this precipitation were random, chaotic and, geologically speaking, very slow. These ideas were, of course, at odds with a literal reading of Genesis that seemed to indicate an *ex nihilo* creation of solid rock in six days. The contrary nature of these two different interpretations is the context within which Penn wrote his book. Penn wryly addresses this issue by essentially pitting the mineral geologists against none other than Isaac Newton. Newton had written on this very subject and Penn quotes him at large:

---

2 Scriptural geologists wrote in the 19th century and were a diverse group of scholars, scientists and/or ministers who believed in the literal interpretation of Genesis. They were much like modern YEC, except they tended to accept the science of the day without too much reservation (unlike many modern YEC).

> *It seems probable to me (said the wise, sober, and circumspect Newton) that God, in the beginning, formed matter, in solid, massy, hard, impenetrable, movable particles, of such sizes and figures, and with such other properties, and in such proportions to space, as most conduced to the end for which He formed them.— All the material things seem to have been composed of the hard and solid particles above mentioned, variously associated in the first creation by the counsels of an Intelligent Agent. For it became Him who created them to set them in order; and if He did so, it is unphilosophical to seek for any other origin of this world, or to pretend that it might rise out of a chaos by the mere laws of Nature; though, being once formed, it may continue by those laws for many ages (Penn 1825, p. 28).*

Newton clearly accepted a literal reading of Genesis,[3] and as such, sought to interpret the origin of the solid Earth in terms of a supernatural event. Although Newton did not use the term "mature creation," it would seem natural to assume that he was conceptually thinking along those lines.

Penn picks up on Newton's line of reasoning and uses it to set the stage for a formal mature creation apologetic. Penn calls those things brought into existence during Creation Week "first formations." According to Penn, these "first formations," by virtue of their being created "first," must be excluded from the natural processes that precede their formation in the present. Humans, animals and plants, for example, are not "born" mature organisms. They begin their cycle of life as single cells that grow and multiply over time.

---

3 Although he may have changed his position later in life (Millam 2011).

The first man, Adam, was not "grown" from a single cell. He was created as a fully grown adult:[4] "Now, whatever be the standard of age in correspondence to which we may suppose that first man to have been created, it must correspond to some period of the human life subsequent to the birth" (Penn 1825, p. 60).

Dissenters of the traditional interpretation of Genesis, although aligning with more modern concepts in geology, nevertheless still adhered to a literal and supernatural creation of Adam and Eve. Penn's logic was thus founded on reasonable and rational grounds, at least at that time. From this foundation, Penn proceeds to draw out reasonable inferences:

> *If a bone of that first, created man [Adam] now remained, and were mingled with other bones, pertaining to a generated race [people born after Adam]; and if it were to be submitted to the inspection and examination of an anatomist, what opinion and judgment would its sensible phenomena suggest, respecting the mode of its first formation, and what would be his conclusion? ... He would naturally pronounce of this bone, as of all the other bones; that its fibers were originally soft; until, in the shelter of the maternal womb, it acquired the hardness of a cartilage, and then of bone (Penn 1825, p. 62 – 63).*

Penn draws the following astounding conclusion from these observations:

4 Of course, this is a presupposition. It could be that Adam was created as a boy or a baby and then grown into a mature man. I am going to take the position that he was created as a mature man.

> *... from hence we obtain a second principle, with respect to such first formations by creation: That their sensible phenomena alone* ***cannot determine the mode of their formation****, since the real mode was in direct contradiction to the sensible indications of those phenomena (Penn 1825, p. 64; emphasis mine).*

And this:

> *His first formations were made in correspondence with the laws which He was then about to establish, and in anticipation of ... appearances which were thenceforward to be produced only by the operations of those laws (Penn 1825, p. 77).*

According to Penn, one must believe what God said about the genesis of Adam's bones, *despite* what the anatomical (perhaps today he would have used the term "osteological"[5]) evidence might suggest. Anatomically and osteologically, Adam's bones would look no different from any other man's bones. That is because Adam's bones were made "in anticipation of ... [soon to be established natural] laws." Those future laws would shape men's bones into the mirror image of the first created bones because God anticipated their appearance and composition before they would ever even grow to maturity.

Penn convincingly moves from these straightforward biological observations to those representative of geology—namely, the "first formations" of the Earth's crust. He uses the same line of reasoning to suggest that "first formations" of inorganic minerals anticipated the appearance and composition of their "naturally" formed counterparts (Penn 1825, p. 71). This leads Penn to assert that "that primitive

5 The study of bones.

immediate crystallization, can furnish no data for computing time, more than primitive immediate ossification, or primitive immediate lignification" (Penn 1825, p. 72). In other words, although Adam's bones and Eden's trees would suggest processes leading to real histories, those processes and histories, as with those of the crust's first crystals, are only *apparent*.

The second most important mature creation apologetic was published in 1857 by Phillip Henry Gosse, a well-known and respected ornithologist. In his book, *Omphalos: An Attempt to Untie the Geological Knot*, Gosse suggests that the entire fossil record was specially created by God during Creation Week. Gosse uses the analogy of cycles.

The life cycle of a plant, for example, does not have a beginning or an end. During Creation Week, God could have brought plants into being at any stage in their life cycle since there is no legitimate *beginning* to their existence. Accordingly, since plants die, this state should also be incorporated into the cycle. Death is not the end of the cycle; it is merely a *part of it*. God could choose, therefore, to bring a plant into being at any stage of its existence—as a seed, a seedling, a mature plant or a dead plant.

These cycles are also true for all living things, including animals. According to Gosse, the entire fossil record represents billions of organisms that were created in the "death stage" rather than in the seedling, newborn or even adult stage. This means that fossils would *appear* to have had a life cycle and thus a history *prior* to their being created. Theoretically, Gosse is right. And, in fact, God did something like this when He created Adam. Adam was not created as a newborn baby, a young boy or an old man, although he *could* have been. Rather, Adam was created as a sexually mature man.

But is it feasible for God to bring organisms into existence at the stage of their death? I suppose He could have, but to do so would conflict with God's creative strategy. God created a mature universe in preparation for His crowning achievement—man. Developing a functional universe in six days *suits* this creative strategy. There is nothing strategic about creating a vast record of dead animals. Such a record serves no creative or teleological purpose. On the contrary, given God's initial design that seems to exclude suffering and death, such a record would serve to *condone* suffering and death. The Bible teaches that animal death occurred only *after* the fall of Adam.[6]

---

6 Some Christians see no problem with animal death prior to the Fall of Adam. It is true, the Bible does not *explicitly* speak of whether or not animals died before the Fall, but I think it does so *implicitly*. Consider Genesis 1:29 – 30 (ESV) and God's prescriptive design for human and animal diets: "And God said, 'Behold, I have given you every plant yielding seed that is on the face of all the earth, and every tree with seed in its fruit. You shall have them for food. And to every beast of the earth and to every bird of the heavens and to everything that creeps on the earth, everything that has the breath of life, I have given every green plant for food.'" Some (Ross 1994; Poythress 2019) contend that although this verse prescribes a particular kind of diet, it does not *forbid* others (eating of meat). I see the logic behind this possibility, but I think it's a stretch. Does it not seem odd for God to discuss the diet of herbivores but remain completely silent on the most fundamental aspect of carnivory—the food chain?

Secondly, the Curse itself extended to the animal realm (Genesis 3:14). If animal death, carnivory, and suffering are not related to the Curse, then in what sense were they affected?

Thirdly, the association of carnivory with the Curse seems to be picked up by later Bible writers as they depict a future time when carnivores will live in peace with their prey (Isaiah 11:6).

Finally, I think the idea of animal suffering in general to be antithetical to a "good" creation. I will recount a descriptive, if not awful, example narrated by Kurt Wise at a recent conference (Origins 2019 in Cedarville, Ohio). Scientists were recently able to detect the last moments in the life of an Ice Age mammoth. This mammoth became stuck in the mud. In order to escape from the mud, the mammoth resorted to violent twisting, which unfortunately resulted in spiral fractures to all four of its limbs. The mammoth, found in an upright position, eventually died from blood poisoning. This means that over the many days it took for the blood poisoning to

This means that any history, or *apparent* history, predating Adam cannot include real, or *apparent*, animal deaths.

A contemporary of Gosse, known only as C.B., wrote *Geology in its Relation to Revealed Religion* (C.B. 1853). C.B. does not articulate a mature creation apologetic per se, but he does suggest, along with Gosse (1857) and Penn (1825), that the appearance of maturity or age is a necessary function of ex nihilo creation:

> *If we trace the existence of plants or animals through a series of generations, we are justified in deducing analogical conclusions, but not so when the first plants or animals, as they came from the hands of the Creator, are the subject of our consideration. In them, we would, too, discover every trace of a former existence which they never passed. Here our analogy, though true in appearance, would be false in effect (C.B. 1853, p. v).*

---

finally kill the animal, it writhed in pain as it bore its weight on four broken limbs. What a horrible account of suffering. Something seems very wrong with this picture. The fact that I am inwardly disturbed by this story must surely preclude its being called "very good."

When defending carnivory, many OEC often quote texts that imply God's *original* meat-eating design (Poythress 2019, p. 159). Psalm 104:21 (ESV) is a good example: "The young lions roar for their prey, seeking their food from God." The problem with using these texts is that they do not distinguish between God's design in nature *prior* to the Fall versus His design in nature *subsequent* to the Fall. That God designed the Lion with teeth, agility and instinct for the purpose of killing other animals is not questioned. The question is, *when* did these characteristics get "turned on"? Using texts like Psalm 104 is of no help because they were written by prophets living in a post-Fall world. Consider verse 35 of the *same* Psalm: "Let sinners be consumed from the earth, and let the wicked be no more!" (Psalm 104:35 ESV). Surely, the writer is not thinking of humans living in an Edenic state? More than likely, Psalm 104 incorporates eclectic facets from the Creation Week, the Flood narrative, as well as life in general. Using these texts to support animal death *prior* to the Fall is, therefore, unsubstantiated.

Along with Penn, C.B. contends that the primary or crystalline granites found beneath the sedimentary or secondary rocks were created during Creation Week, but he goes further by admitting the same origin for coal and even limestone (C.B. p. 125).

C.B.'s work is better understood in terms of a scientific point-by-point rebuttal against the growing scientific belief in the antiquity of the Earth, rather than an innovative, philosophical discussion such as that seen in the works of Penn (1825) and Gosse (1857). Of the latter three works, only Penn's (1825) remains both faithful to the biblical text, while at the same time cohesively engaging the science of his day. C.B.'s (1853) work does not wrestle with seemingly solid data, but instead dismisses the evidence by appealing to credulity.

Numerous fossil "shrimps,""footprints" and other trace fossils, for example, are removed from the realm of ancient biology with a wave of the hand and an appeal to their two dimensional character: "We cannot imagine these impressions were made by the bodies of actual animals ... they are simple surfaces, and yet, they are the remains of animals! ... who could believe such absurdities?" (C.B. 1853, p. 150). C.B. instead suggests that these body, mold and ichnofossils are best explained as examples of apophenia:[7]

> *Marks of this kind are so common, and assume such varieties of form, that they must sometimes represent objects with which we are familiar. Nay, the human mind is prone to group together the most dissimilar objects, and form them into figures; and hence, human faces and human figures, without limit, animals, trees,*

7 This is the human tendency to see faces in clouds or other meaningful images in abstract objects.

> *&c., are seen by those who pore over a fire, the clouds, variegated marble, &c (C.B. 1853, p. 150 – 151).*

As for Gosse, his ideas and his book were not taken seriously by the scientists or the theologians of his day. The reason for this most likely stems from what Craig Rusbult (2003) calls "Total Apparent Age (TAA)." This is the application of a mature creation apologetic to almost *every* geological, astronomical, paleontological and biological event in history. Taken to its logical conclusion, TAA could even include all historical events and every thought leading up to those that occurred just five minutes ago. In other words, the entirety of one's existence is only *apparent*. Such thinking has prompted some writers to conclude, "... if we accept the concept of 'apparent age,' we are left with no assurance of the reality of any history whatever" (Menninga 1988).

The 20th century

Most modern YEC reject Gosse's approach opting for what Rusbult (2003) calls Partial Apparent Age (PAA) or Minimal Apparent Age (MAA). PAA essentially accepts Gosse's argument for the universe, the Earth, and the life forms created *during* Creation Week but does not include any events *subsequent* to Day 6 of Creation Week. Noah's Flood, for example, could not be included within a PAA model of origins.

Henry Morris was the most well-known modern YEC holding to PAA (Rusbult 2003, part II). In his writings, Morris emphasizes the importance of mature creation apologetics, but only for creative events that occurred during Creation Week. Morris argues that starlight, for example, was created *in-transit*, entirely separate from the luminaries of origin:

> *The light from the sun, moon, and stars was shining upon the earth as soon as they were created, since their very purpose was 'to give light upon the earth' (Genesis 1:17). As a matter of fact, it is possible that these light-waves were energized even before the heavenly bodies themselves in order to provide the light for the first three days (Morris 1974, p. 210).*

MAA proponents like the idea of PAA but do not like the full ramifications of PAA. For example, SN1987A, an exploding supernova that was first observed by humans in 1987, represents an event that, according to the secular paradigm, occurred about 170,000 years ago. According to PAA proponents, this event could not have really occurred because the Earth is only 6,000 to 10,000 years old. This means that the neutrinos, visible light, X-rays, and gamma-rays that represent the supernova were real, but the actual exploding star was not. These particles were merely created by God *in-transit*. Many MAA proponents believe that creative acts, such as SN1987A, deserve a *real* history, and as such seek to explain these more complicated phenomena using YEC cosmological models, such as exponentially-decaying light speed, curved space, and large-scale relativistic time-dilation, among others.

The 21st century

In the last two decades, YEC cosmological models have somewhat displaced mature creation apologetics in general. Most YEC still insist on a mature creation including a fully functioning biosphere (for example, Adam and Eve), but tend to shy away from deeper, more intricate explanations when it comes to cosmological phenomena. This tendency for some YEC to mitigate the importance of a generalized form of mature creation apologetics is often centered around the meaning of the word "age" from the phrase, "apparent or appearance of age" (Lisle 2015, p. 291).

Although other well-known YEC have taken this stance, I will quote from Dr. Jason Lisle in particular. Now, I want to say at the outset that I wholeheartedly respect Dr. Jason Lisle and fully support the important contribution he is making to YEC research. The comments that follow are related only to the way in which we both differ as to the place of mature creation explanations in modern YEC apologetics.

According to Lisle (2015 p. 291), "age" cannot have an "appearance." Lisle rightly contends that the concept of age is something each of us learns through experience. He uses the Earth as an example: "but for the earth, there is no sample of planets of multiple ages for which to make the comparison" (Lisle 2015, p. 292). In other words, humans have never seen planets mature and grow over time. As such, we have no *experiential* foundation from which to judge the age of the Earth. He goes on to say: "If people think the earth 'looks old,' this is due to unbiblical thinking on their part, not any deception on God's part" (Lisle 2015, p. 293). Accordingly, we should not say the Earth *looks old*. If we mean by this that the Earth *looks* old because we can somehow compare it to other planets, then I wholeheartedly agree with Lisle. Unfortunately, Lisle's reasoning does not appreciate the full ramifications of what is at stake. The book I have quoted from (Lisle 2015) is a concept-by-concept rebuttal of Dr. Hugh Ross' old-Earth creationism. Since Ross consistently applies radioisotope chronologies to the age of the Earth and the universe, he obviously uses big numbers consistent with radioisotope dating. Billions of years and the use of numbers such as 4.6 billion years (the secular age of the Earth) or 14 billion years (the secular age of the universe) are normative.

The issue for many OEC, however, is not necessarily the radio-isotopic age of the Earth. This is a side issue. The problem is the vast amount of time required for large structures like the Earth to form from geophysical processes such as magma-mixing, fractionation, partial melting and gravitational differentiation[8]—processes that are *currently in operation* (Deem 2013). John Baumgardner, a prominent and highly respected YEC clearly believes that these geophysical processes were at work during the formation of the Earth (Baumgardner 2000, p. 50).

A big number, like 4.6 billion years, is irrelevant to any discussion on the age of the Earth because, quite simply, it is not about the *age*. *Current processes* are all that is required to perceive that this planet was not formed in a *single day, week, year or even a thousand years*. The issue then is not the semantics associated with the word "age," it is the measurable processes at work within the Earth that, when extrapolated backwards, logically require millions or billions of years to form a mature planet.

Now, I know Lisle understands this, but by missing the *process* aspect for the sake of disparaging the *age* aspect, he has inadvertently put Ross and other OEC offside. Creationists should be *agreeing* with Ross about our highly developed planet, not disagreeing. Comments such as this by Lisle: "It is an egregious fallacy ... to argue that the earth must be old since it 'looks old,' because, frankly, it doesn't" (Lisle 2015, p. 292 – 293), fail to appreciate the *heart* of the problem. This is not what our OEC brothers and sisters need to hear. They need to be assured that the Earth does *seem* to have a long history

8 Some of these processes are explained in Chapters 4 and 5.

associated with it, as does the universe, but that God has not left us without His Word that clearly informs us of their *actual* history. Notice that the conflict has now been removed from the realm of science to the realm of *revelation*.

Common objections to mature creation apologetics

As stated above, there are many objections to mature creation apologetics, but one deserves a little more attention. Rusbult (2003, part II), has written an interesting article that makes a distinction between what he calls Essential Apparent Age (EAA) and Non-Essential Apparent Age (NEAA).

EAA underscores an apparent age that is *essential* for an instantaneously functioning, supernaturally created universe. For example, soil—the product of weathering over time and a necessary component of a life-sustaining, fully functioning ecosystem—must have been created *mature*. A solid, crystalline crust, streams and valleys, and fully grown mountains would also be essential to a functioning, supernaturally created ecosystem. Even light from distant stars falls into this category since the Bible states that it had an immediate purpose (Genesis 1:14 – 19). Philosophically speaking, critics of mature creationism should at least be able to see the logic behind EAA.

Problems arise, however, with NEAA. SN1987A is an excellent example of the latter (discussed above). Why would God create *in-transit* starlight representative of a fake supernova that itself becomes manifest only thousands of years later, and then only for a few months before fading into the stellar background? It seems fair to say that SN1987A and its related cosmological phenomena do not seem to be required for the immediate functionality of a mature universe (unpacked further in Chapter 6).

A biological example that deals specifically with the supernatural creation of Adam and Eve is discussed by Lindsay Harold from The Creation Club (2017). Although Harold does not use the terms EAA and NEAA, her overall thesis draws upon this distinction. At the outset, I want to commend Harold for dealing with an extremely tricky topic. She endeavors to wrestle with the data to provide an overall picture that avoids pseudo-science while at the same time remaining faithful to scientific methodology. I have quoted extensively from her to provide a context for this issue:

> *In some [YE] creationist circles, because we believe in a supernatural creation that was made mature, and thus that some things may look older than they actually are, some may wish to simply accept all cases of apparent age as being created that way. This, however, is a bad approach. It's a God-of-the-gaps argument in which we simply assert that God created it that way to avoid having to deal with the evidence. And when we use that sort of argument, we give ammunition to those who wish to criticize Biblical creation ... For example, while Adam was created as an adult, and adults usually have to go through the childhood stages to reach adulthood, Adam didn't. He was created mature. However, he was not created with false signs of aging. Thus, on first glance, a doctor examining him right after his creation would have thought he was 20 or 30 years old due to the body size, ability to grow a beard, sexual maturity, etc. But on closer inspection, the doctor would notice something peculiar. There would be no wrinkles, no scars, no callouses, and no belly button. Adam wouldn't show signs of actual aging or having lived years on the earth or having been injured in events that never happened. If he had, that would be apparent age, not merely a mature*

> *creation. Close inspection would be needed to indicate the difference. Whereas a cursory inspection and initial conclusion might point to a man who had lived a couple of decades or more, closer and more detailed inspection would give reason to think he hadn't actually started as a baby and grown to adulthood, and thus the appearance of age would disappear. God also created animals fully grown, not as embryos, and created trees that were mature and bearing fruit, not mere seeds or seedlings. Looking at a newly created tree, one might think it had taken decades to grow to that size, but closer study might find that the tree has no growth rings indicating seasonal changes it had never experienced or that it has no scars from broken branches that never existed. A newly created animal would similarly bear no scars or defects indicating a history that never occurred.*

Harold proposes a number of interesting phenomena peculiar to the very first organisms created by God: animals lacked scars and defects, trees lacked rings, and Adam lacked a belly button, callouses and even age-related wrinkles normative for a man of about 20 to 30 years of age. According to Harold, to give Adam a belly button, trees rings and animals defects would amount to deception on God's part in what she calls a "false appearance of age."

Rusbult (2003, part II) would describe such phenomena in terms of Non-Essential Apparent Age. Addressing these legitimate concerns will be pertinent to a thorough mature creation apologetic. First, it will be important to compartmentalize these various phenomena.

It is true that Adam and the original animals would not have had scars or other kinds of defects, but that is because they were created as perfect, unfallen creatures from the very beginning. It would seem unusual, then, for God to create Adam with a scar or an animal with a defect. But what about signs of aging? Furnishing Adam with age-related features does not necessarily imply deception. Human maturation is a natural process designed to unfold over time under specific environmental conditions. Since Adam was created *anticipating* those environmental conditions, he may have been created *as though* he had been exposed to them.

Consider Adam's skin. Skin requires regular bouts of sunlight to protect it from burning (Seigel 2014). Rather than create Adam with *perfect* skin, it may be that God "exposed" Adam to "sunlight" so that his skin developed to the same extent as other 20 to 30 year-old men. This exposure may have left his skin less than perfect (whatever that means anyway), having darker spots, blemishes and even wrinkles. Similarly, Adam's feet were more than likely calloused for the purpose of protecting his feet, even though he had never taken a step prior to his creation.

What about defects, scars, and signs of aging for similar acts of special creation *post*-Fall? Consider Moses' snake (Exodus 4:3). Did God make a genetically perfect snake without defect and without any signs of aging? If, for example, Moses' snake had been a rattlesnake (just humor me), and if God had not customized the snake with elements of maturity then it would have had only a nub for its rattle. As a rattlesnake matures, it adds rattles to its congenital nub (Steen 2011). What about scars and other signs of age-related wear and tear? It would have been rather strange to Moses' Egyptian opponents, who would no doubt be very familiar with this kind of snake (whatever it was), to see a physically perfect adult rattlesnake wearing nothing but its birthday suit and shaking its rather pathetic booty. It makes sense that the snake God created would have had about the same number of genetic mutations as other snakes living at that time, would exhibit age-related phenomena consistent with mature rattlesnakes, and would also be furnished with phenomena consistent with a fallen world.

I do not think it would have been deceptive of God to create such a snake. God's *purpose* in creating Moses' snake was to convince his naysayers that God is none other than the Creator. A pathetic knock-off was just not going to cut it. These same criteria would have been true for everything God supernaturally created *post*-Fall. This includes the wine, the bread and the fish that Jesus brought into existence in the gospels. These objects would possess all the characteristics common to those from which they were replicated, warts and all.

What about rings in the first of Eden's trees? The problem here is not as simple as it first seems. It turns out that many living organisms have some kind of growth ring. Shelled organisms are a good example. Certain shelled organisms have yearly "rings" consistent with periodic, environmentally induced calcium deficiencies in the water (Jones 1989). Similar phenomena are consistent in the bones and teeth of most vertebrates. Lines of arrested growth (LAGs) are widely used by paleontologists to estimate, for example, the ontogenetic development of dinosaurs. According to Fastovsky and Weishampel (2012 p. 275), these LAGs can even be influenced by seasonal and/or annual cycles.

Many such growth-related characteristics have more than a simple, age-related purpose. In trees, for example, the denser, hard or late-wood rings of a tree are essential for strength. Without this particular ring, the tree would collapse under its own weight (Wimmer 2010). Although no expert, I suggest that shelled organisms and vertebrate bones most likely have similar constraints.

An across-the-board, blanket-like removal of all such age and growth-related phenomena by God during Creation Week would change the very way nature was supposed to look, grow, feel and sound (rattlesnakes!). Such an anomaly would also make it difficult, if not impossible, for man to complete his God-given mandate to have dominion over the entire Earth. To have dominion over something means to *understand* how it works. If Adam suddenly decided to dump his newly acquired theological training in exchange for a career in paleontology, archeology or osteology, he would soon find himself up against some contradictory and anachronistic data.

If we assume that God created the Earth and its biosphere in *anticipation* of this mandate, and in anticipation of the normal laws of the universe, then it was not deceptive for God to give trees rings, Adam wrinkles, and Eve a belly button. A direct application of these observations translates to a similar set of processes at work for the inorganic creation—rocks, mountains, seas, streams, landforms, continents, soil, the crust, and, in fact, the entire planet.

What's in a name?

Misconceptions within language commonly occur because of ill-defined terms and/or preconceived ideas. Terms like "mature creation," "appearance of age" and/or "apparent history," as we have seen, present the audience with nuances of meaning that can sometimes be confusing. Over the past few decades, this confusion has often cast these terms in a negative light, causing some Christians to doubt the usefulness of what I think is an *explicitly* biblical doctrine. The major problem with these terms, however, is the *static* language used to describe *dynamic* events. It is almost as if the *cause* is missing from the *effect.*

Consider, for example, the term "appearance of age." Many understand this term to imply that God "antiqued" the creation, giving it an apparent "age," for no other reason other than He wanted it to *look* old (like a new pair of jeans that have been distressed). Yet this is not what the term means. The Earth, for example, looks old, *not* because God is into the distressed look. The Earth looks old because God used large-scale processes to bring it to maturity. These large-scale processes are then *interpreted* by humans as the passage of time. This distinction makes all the difference. The Earth looks old then, *not* because God wanted it to look old, but because of the authority we place upon *our* experiences.

Given all of this, I would like to introduce *dynamic* terminology that underscores the importance of *processes.*

Supernatural formative processes (SFP) are those that God used to specially create and/or alter normal, everyday objects. Turning regular water into drinkable wine (John 2:1 – 11), restoring a withered arm (Mark 3:47), multiplying bread and fish (Mark 6:30 – 44), creating Adam from the dust (Genesis 2:7), causing the rapid growth of the biosphere (Genesis 1:11) or the rapid death of a fig tree (Matthew 21:18 – 22) are all examples of SFP.

As with many aspects of supernatural metaphysics, there is no intuitive rationale that can easily explain the mechanisms used. Why, for example, did Jesus ask the servants to fill the jars with water (John 2:7)? Why not just create the wine *ex nihilo*? Similarly, why did Jesus acquire bread and fish before feeding the multitudes (Mark 6:41)? Other instances, although similar in some respects, are very different in others. In Matthew 21:18 – 22, Jesus caused a fig tree to wither and die overnight. The exact opposite occurred in Genesis 1:11 when God commanded the biosphere to sprout and mature in a single afternoon. There are also many biological examples that have their own metaphysical nuances. For example, in John 9, a man's long dead and atrophied retinal cells are regenerated, and in Acts 3:6 emaciated limbs grow new muscle, tendons and ligaments.

Deriving a common set of principles from these examples may at first seem challenging, but there are a few factors that unite them.

First, excluding the events of Creation Week, God did not specially create anything innovative or novel. Supernaturally created and/or altered objects were always patterned on *then-existing natural* designs. God did not, for example, create unknown natural substances (perhaps found elsewhere in the universe), electrical devices like laptops (found only in the present), or long-extinct animals (found only in the past).

Second, SFP always resulted in objects that *retained natural histories.* The newly created and/or altered objects were very ordinary, and most likely looked like objects that were quite familiar to that time and place.

Third, the formation of created and/or altered objects occurred without time restraints. In every Scriptural example, the object was formed in an instant (wine), or restored very, very rapidly (a withered arm). Importantly, these rates of change were not accelerated by means of *natural* mechanisms, such as an enzyme that speeds up cell metabolism or a flood that rapidly erodes rock, but were clearly *supernatural*, as in a fig tree that withered overnight or a forest that "sprouted" up in an afternoon.

In Chapter 4, I will demonstrate that in every one of these examples, God's supernatural creative strategy involved a *commitment* to *existing natural processes.* In Chapters 5 and 6, I will demonstrate that this commitment extended to the creation of the Earth, and perhaps even to the entire universe.

Chapter 3

Biblical Examples of Supernatural Formative Processes (SFP)

Examples of SFP found *outside* of the book of Genesis

Jesus turns water into mature wine

In John 2:3, Jesus' mother asks Jesus to provide more wine for the feast. We are unsure whether Mary intended Jesus to supernaturally create the wine, or if she simply wanted Jesus to physically go and pick it up from the corner store. That Jesus indicated his "hour had not yet come" seems to point in the direction of the former. Whatever her motive, Jesus decided to use this occasion to "manifest his glory" to his disciples (John 2:11 ESV), and so supernaturally turned ordinary water into mature wine.

Of interest to us are the scientific/theological ramifications of this miracle. If we could somehow procure a sample of that wine and submit it to a series of experiments in a modern scientific laboratory, what kinds of results might we expect as to the wine's origin? Without revealing its true provenance to a secular scientist, he may conclude that the grapes had been grown in a particular climate, at a particular geographical location, and under the stewardship of a particular winemaker over a certain period of extended time. At the very least, he would believe that a vine had been grown, grapes had been crushed, and juice had been fermented. In other words, he would conclude that the wine retained a *natural history*. Of course, this conclusion would be at odds with what actually occurred, especially given the time involved.

Mature almonds bud from Aaron's staff

In Numbers 17, each chief from each of the 12 tribes of Israel was required to deposit his staff in the tent of meeting. God would show these men that He had sovereignly chosen Aaron by supernaturally causing his wooden staff to blossom and produce a mature crop of almonds overnight. Yet almond trees require a long period of immature incubation—upwards of five years—before a viable and economically mature crop of almonds can be produced. The origin of this crop of almonds is, of course, beyond the scope of normal science. Yet if this small crop of almonds were harvested and sent to a modern laboratory, what kind of predictable, historical conclusions might one expect to find? Like the wine, the scientist would conclude that the almonds had grown using a natural set of processes. And in fact, we are told that the staff, "had *sprouted* and *put forth* buds and *produced* blossoms, and it *bore* ripe almonds" (Numbers 17: 8b ESV). In other words, the growth and maturation of the almonds mirrored *naturalistic processes*.

A stick is turned into a snake

One of the most interesting examples of SFP involves God turning Moses' stick into a living, breathing snake (Exodus 4:3). It is almost certain that the snake had a complete skeleton, brain, lungs, liver, heart, stomach, spleen, pancreas, testes, small intestine, kidneys, and rectum, all of which, from our perspective, must have developed from the blueprint found in the snake's DNA. A herpetologist would likely assume that the snake had an embryonic history, which itself demands the existence of haploid gametes that came from separate male and female snakes. The herpetologist would, of course, be in error, since the snake's natural history *must* be attributed to the work of SFP. This example is most likely representative of Creation Week when God created fully formed, mature organisms.

Jesus' DNA

An example that to my knowledge has never been considered elsewhere is found in the miracle of the incarnation and the biological constituents of Jesus' DNA. Jesus is fully God, but he is also fully human:

> *For this reason he had to be made like them,* ***fully human in every way****, in order that he might become a merciful and faithful high priest in service to God, and that he might make atonement for the sins of the people (Hebrews 2:17 NIV; emphasis mine).*

Jesus was fully human "in every way." We should assume, therefore, that Jesus' biology was also fully human. This would mean that Jesus had diploid somatic cells, just like all other humans. All humans possess DNA inherited from the father *and* mother. We call these kinds of cells *diploid*, meaning "two sets." This means that all humans have one set of chromosomes that came from the father's sperm and one set of chromosomes that came from the mother's ovum. This biological characteristic makes us distinctly human.

Jesus would not be truly human without this biological characteristic. The fact that Jesus' somatic (body) cells are diploid means, of course, that the constituents of Jesus' biological makeup resulted from an extraordinary miracle well beyond our understanding. But consider the philosophical and scientific implications of such a miracle. If a biologist could run a test on Jesus' blood, the results would almost certainly indicate a chromosomal blueprint that was biologically indistinguishable from any other human. If the biologist was an unbeliever and if the blood's provenance was not disclosed, the biologist would assume that the blood's owner was nothing more than a man with a biological past just like everybody else.

Indeed, reaching this conclusion is not dependent on modern scientific techniques. Jesus' own people thought he was a rather unremarkable man: "Is not this the carpenter's son? Is not his mother called Mary? And are not his brothers James and Joseph and Simon and Judas? And are not all his sisters with us?" (Mathew 13:55 – 56 ESV). In summary, Jesus' biology displays clear aspects of SFP, whereby one set of chromosomes was created to mirror a fully diverse set of *naturally* formed male chromosomes.

Many other examples of SFP can be found in the Bible. Jesus creates bread and fish on multiple occasions (Example: Matthew 14:13 – 21) and atrophied bones, muscles and organs are restored to the extent that modern doctors would not detect their original condition (John 5:9; John 9:6).

Abstract examples of SFP

Time-dependent *abstract* examples also exist. In Acts 3:1 – 10, for example, we are told that a man lame from birth is healed and thereafter able to walk. Yet this is a double miracle. The man was *born* lame. The ability to walk is both physical and *learned.* We all know that it takes many years for toddlers to master this ability, and much the same time frame is required for adults who have had to relearn to walk after an accident. Yet this man in Acts was not only healed, he was somehow infused with a *time-dependent, learned* skill. Adam and Eve must have been infused with similar learned skills.

Examples of SFP from the Genesis account

The creation account itself presupposes a commitment to SFP for the entire animal kingdom, something made more explicit in the creation of Adam and Eve.

YEC Kurt Wise (2002) indicates that animals were not just created mature but were created within ecosystems that consisted of complete age distributions—babies, fledglings, and adults. He argues similarly for the creation of the plant world. The reasoning for this age distribution stems from a reasonable assumption of survivability. Many ecosystems *require* such an environment for survival, and so they must have been created with an age distribution from the very beginning (Wise 2002).

This is an interesting suggestion because it aligns itself with one of the main features foundational to the thesis of this book—that is, that God seems to have *anticipated* future *sustaining* processes even before applying SFP. This idea is not novel. Isaac Newton came to this conclusion over 400 years ago:

> *It seems probable to me ... that God, in the beginning, formed matter, in solid, massy, hard, impenetrable, movable particles, of such sizes and figures, and with such other properties, and in such proportions to space,* ***as most conduced to the end for which He formed them*** *(Penn 1825, p. 28; emphasis mine).*

The importance of *process* to SFP

Although the idea of *process* is somewhat ambiguous in some of the above examples, there are others that seem to biblically require it. The vegetated world, for example, is said to have been "brought forth" (Genesis 1:11 – 12 ESV) in what appears to be an unfolding process rather than an instantaneous creation.

Rapidly unfolding processes seem to be required for other events. YEC Andrew Snelling (2008), for example, suggests that deposition of kilometer-thick strata must have been achieved in just a few hours at today's rates. If God brought forth the land in just a single day, then erosion must have been extremely rapid and intense. Yet at today's rates it would take tens or hundreds of years for the finest of those sediments to settle out of the water column. How would the first organisms have lived in seas filled with muddy soup? How would photosynthesis have been possible? It would seem, then, that supernaturally rapid erosion was also followed by supernaturally rapid deposition.

YEC Danny Faulkner (2014) follows Snelling's (2008) commitment to accelerated processes as he discusses rapid cratering for the moon on Day 4 (Faulkner 2014).

Thinking through such examples implies not just *process*, and not just *accelerated process*, but also seems to require the *constancy* of *relative rates of process* (unpacked in Chapter 4).

Following the scientific evidence always leads to the truth, or does it?

Scientism is the belief that truth about the physical universe can be found only by carefully applying the scientific method to natural phenomena.[1] Absolute truth about the physical universe cannot

1 See Moreland (2018) for an excellent discussion on this.

be found in any other discipline, including religion. Yet is this assumption *itself* true?

Consider the examples of SFP discussed in this chapter that exist *outside* of the book of Genesis. In all of these, God created objects that were patterned on then-present analogies, things that you would find or use in everyday life—wine, bread, fish, etc. A commitment to scientism would lead an investigator to conclude that these things, as well as Aaron's almonds, Moses' snake, and even Jesus' DNA, all had very real histories because experience teaches us that time-dependent process is *required* for these things to come into existence.

Yet clearly, following the evidence in each of these cases would lead to erroneous conclusions. This is *key* for all that follows in this book. When God says that He has acted in the natural realm to create objects supernaturally, humans are not given the option to interpret their origin in terms of a materialistic worldview. Moreover, if humans decide to forego God's revelatory interpretation, they will undoubtedly discover a *false* origin that *will* lead to erroneous conclusions.

This is in direct contradiction to Ross (1994): "Therefore, no contradiction between the facts of nature and the facts of the Bible is possible. Any apparent contradiction must stem from human misunderstanding."

Jesus' wine, bread, fish, and even DNA would *only* furnish an interpretation contrary to that procured using scientific methodology. The *real* history for each of the latter examples can be ascertained *only* by believing what God has said about them.

Objections to *biblical* examples of SFP

Interestingly, some OEC deny the implications for some of these biblical examples. Regarding Adam and Jesus' wine, Richard Deem (2013) says:

> *Many young-Earth creationists claim that the universe and earth just appear to be old, but are really young. Frequently cited to support such a view are scriptures that indicate God created Adam as an adult, and Jesus created wine that appeared to be aged. According to appearance of age advocates, God created Adam with the appearance of age, since he seems to have been created as an adult. The Bible says that Jesus turned water into wine at the wedding in Cana. Didn't this wine have the appearance of age? There are two major problems with these analogies. First, neither the body of Adam nor the wine of Cana is available for inspection. The creation is still here.*

Ross (1994) uses a similar argument: "Neither Adam and Eve nor the water Jesus turned into wine are available for accurate age testing. Adam and Eve died. The wine was consumed. Therefore, neither should be brought forward as evidence supporting appearance of age."

I find this line of reasoning inconsistent. Neither Deem nor Ross has ever seen the resurrected Jesus, yet I have no doubt that they believe Jesus' disfigured body, complete with internal organ failure (by piercing) and at least a few days' worth of decomposition, was not only miraculously restored, but was transformed into a body fit for eternity. Using their own objections, I could say that Deem and Ross should not believe that Jesus died and was subsequently resurrected unless they can somehow gain access to Jesus' body and so validate these claims scientifically.

We may not have Jesus' wine, but we do have a very clear testimony as to the *genuineness* of that wine, as we do to the *genuineness* of Jesus' death and resurrection. This genuineness is in the revelation of the Word of God itself. Since Deem and Ross fully accept the death and resurrection of Jesus based solely on eyewitness accounts, they should also fully accept the eyewitness accounts that attest to the *realness* of Adam and the wine. That Jesus' wine is not available for inspection is, therefore, a moot point.

Deem (2013) continues: "Second, the analogies confuse appearance of age with appearance of history. The creation appears to have a long history, not just an appearance of age or maturity [like the wine]." I am, of course, in *complete* agreement with Deem on this point. This confusion has already been discussed above in Chapter 2 and is essentially the foundation for this entire book. It is not about *age*—as though something *looks* young or old. It is about *process*, and process is always interpreted in terms of age. Jesus' wine, then, is a wonderful example not of *appearance of age*, but of process (SFP) and as such represents a perfect analogy for the creation account which, in Deem's (2013) own words, "appears to have a long history, not just an appearance of age or maturity."

These objections, then, are clearly unwarranted. Every one of these examples above clearly showcases a legitimate theological reality—God uses SFP to create objects that themselves always, *without exception*, mirror naturalistic processes and thus retain *natural* histories. Their inclusion in Scripture is, therefore, useful for drawing a one-on-one analogy with the creation event itself.

Chapter 4

SFP and the Importance of *Process*

The "evolving" Earth

Everything is changing. A human starts life as a zygote, then becomes a baby, then an adult, and then the person dies. A rock starts out in an igneous state. Over time it is broken down into sand and mud and deposited in a stream. Later, the sand is lithified (hardened) and becomes a sandstone that eventually is buried and becomes a metamorphosed quartzite. We call this series of cause and effect relationships, *process*. Processes are fundamentally inherent to the natural order, and they always proceed in a forward direction. Most of the time we can predict, to some degree of accuracy, what processes occurred in the past and where certain processes will ultimately lead. We can do this based on the *relationships* we see between different objects that themselves are caught up in the process of change. These processes and corresponding interpretations are fundamental to the nature of science.

One of the problems with classic mature creation arguments is that they make no room for processes that were at work during Creation Week. Many Christians just assume that God spoke, and *poof*, out came a planet or a star or a tree (classic appearance of age and/or mature creation apologetics). The main reason this straightforward approach exists is because most people are not scientists. They are just normal people who are either unaware or uninterested when it comes to how science interfaces with God's supernaturally created universe. As such, they are content to just take the Genesis account at face value, and that is not a bad thing.

But begin to dig into the natural order *as a scientist* who is also a Christian, and things are no longer that simple. As an example, take modern geophysics. In the last 50 years this discipline has revolutionized the way we view our planet and has successfully described the Earth's composition and general structure based *solely* on modern scientific methodology (Baumgardner 2000). This description is grounded on the modern assumption that our planet is evolving. It has also successfully applied that assumption to processes such as fractionation, partial melting and gravitational differentiation.[1] How is it that secular scientists can effectively explain a process that is at work within the mantle, and then use that knowledge to successfully predict an outcome in the crust? This is pertinent to young-Earth creationism because the processes that these geophysicists depend upon for their data were set in motion *prior* to the end of Creation Week. That geophysical processes were indeed integral to Creation Week is discussed by John Baumgardner, a very well-known and respected geophysicist who is also a YEC (Baumgardner 2000, p. 50):

> *How did this prominent aspect of earth's structure [core, mantle and crust] come about? Did God simply speak it and it instantly*

---

1 All these processes are involved in differentiating or separating the Earth into three basic parts: crust, mantle, and core (discussed further below and in later chapters).

> *appeared that way, or did He speak and a process unfolded, albeit rapidly, that yielded this result? The chemical makeup of the earth's rocks seems to be telling a story that* **God used a process** *(emphasis mine).*

This is fascinating. Baumgardner is suggesting that the differentiation of the core, mantle and crust must have occurred in some kind of progressive fashion—starting off, for example, with mantle-like material that "evolved" into crust-like material. This would explain the successful ability of secular scientists to explain how the internal systems of the Earth work. They are simply basing their assumptions on a chain of processes that extend backwards in time. This is thought-provoking because it means that the processes at work during Creation Week somehow carried over into the time/historical context of Day 7 onwards. How else could the science of geophysics work *unless* this was in fact the case?

Processes currently at work within the Earth are exceedingly slow. How can Christians, therefore, account for the *time* involved in differentiating (separating) the Earth into core, mantle, and crust? Take, for example, the processes involved in the creation of the sea floor. The Earth's oceanic crust is composed of very light minerals (like quartz and plagioclase) relative to those that exist in the mantle (like olivine, pyroxene and amphibole). As mantle rock melts, the *first* minerals to melt are those with lower melting points like quartz and plagioclase. Since these new magmas are lighter than the mantle rock from which they came, they tend to "float" upwards. The last minerals to melt from the mantle rock are those with higher melting points like olivine, pyroxene and amphibole. Since these magmas are heavier than the mantle rock from which they came, they tend to sink. This process is called *partial melting* and is one of the major processes that scientists think differentiates (separates) the crust from the mantle.

As these melts continue to "evolve," some rocks tend to get progressively lighter and lighter (having a higher percentage of quartz and plagioclase) and others get heavier and heavier (having a higher percentage of olivine, pyroxene, and amphibole). Eventually, the lighter melts make their way to the surface to become crust, much like a piece of wood makes its way to the surface after being submerged in water. This is how our current ocean basins are formed. In order to produce a much lighter continental crust, these oceanic rocks need to go through another set of complex processes involving the subduction of the ocean floor back into the mantle.

It is important to realize that these processes are *presently* occurring. It is estimated that the Earth is producing oceanic crust at about 20 cubic kilometers per year (Baumgardner 2000). Continental crust accumulates at slower rates but can also be measured. Dhuime et al. (2018) have estimated that during the Phanerozoic Eon, the Earth produced an average of about 4 cubic kilometers per year, while at the same time destroying about half of that as it returned to the mantle via subduction.

Baumgardner (2000) agrees and has estimated net continental crust accumulation at about 1 cubic kilometer per year. By estimating the current volume of the Earth's continental crust (about 7 billion cubic kilometers) and then extrapolating back the *current* rate at which these rocks are being formed, one can easily approximate a time frame for the crust's formation in the billions of years.

Although many assumptions and approximations underlie these figures, there is no reason for Christians to necessarily reject them. The production of oceanic crust at diverging plate boundaries is a presently occurring and easily measured process (Baumgardner 2000; Stern and Scholl 2010; Dhuime et al. 2018).

The production of continental crust is more difficult to quantify, but it is not impossible. Stern and Scholl (2010) used several methods in their attempt to quantify continental crust accumulation. Although two of those methods assume the great antiquity of the Earth, extrapolating the rate of production/destruction back millions of years, they also used a third method that measured rates of recent volcanic eruptions. Certain volcanoes are known for erupting silica-light magmas (having a lot of quartz and plagioclase). By estimating the eruption rates of these light magmas, it is possible to estimate the volume of the hidden pluton (the portion of the magma body that is underground and that will eventually turn into continental crust). Although this latter method is imprecise, it at least provides unconvinced Christians with persuasive evidence in support of continental crust accumulation.

John Baumgardner himself assures us that we can have a measure of confidence in these assessments: "Let me begin by *affirming* the present-day earth structure as *deduced by* [modern, secular] *seismology* as *firm and trustworthy*" (Baumgardner 2000; emphasis mine).

It is important to understand, however, that these processes were not operating over huge timescales of millions or billions of years. Baumgardner (2000, p. 87) suggests that these processes occurred at *accelerated* rates during the six days of creation:

> *This means in my opinion the processes of nuclear transmutation, mantle convection, magma generation and cooling, together with a spectrum of tectonic and geological processes must have unfolded at rates many orders of magnitude faster than we observe today.*

The nature of the creation account itself indicates some kind of accelerated processes. On Day 1, the Earth was in its most basic form *before* being shaped into a geologically mature planet that could sustain life:

> *God said, "Let the waters under the heavens* ***be gathered*** *together into one place, and let the dry land* ***appear****." And it was so. God called the dry land Earth, and the waters that* ***were gathered*** *together he called Seas (Genesis 1:9 – 10 ESV; emphasis mine).*

These processes involved the movement of vast volumes of water and sediment. That the waters were *gathered* into a *single* location seems to suggest a commitment to basic laws of hydrology. Although this example is a little ambiguous in the details, a more biblically explicit example of process is described for the creation of plants on Day 3:

> *And God said, "let the earth* ***sprout*** *vegetation, plants* ***yielding*** *seed, and fruit trees* ***bearing*** *fruit in which is their seed, each according to its kind, on the earth." And it was so. The earth* ***brought forth*** *vegetation, plants* ***yielding*** *seed according to their own kinds, and trees* ***bearing*** *fruit in which is their seed, each according to its kind. And God saw that it was good (Genesis 1:11 – 12 ESV; emphasis mine).*

Commenting on this verse, Old Testament Hebrew scholar Claus Westermann says:

> *"The earth brought forth." The effect of the command* תדשא הארץ דשא*, "Let the earth put forth vegetation," is described in detail in v. 12* with the word* ותוצא*, "the earth brought forth." ... The meaning of "the bringing forth" is primarily:*

> *"Let something which is within come out." The plants are in the earth and the earth lets them come forth (Westermann 1994, p. 125).*

William D. Barrick, an Old Testament Hebrew scholar and YEC says this in his commentary on Genesis 1 – 11:

> *"Let the earth sprout" (תַּדְשֵׁא הָאָרֶץ, tadšē' hā'āreṣ) might indicate God had already created the seeds for all these plants. He might have done so at the very moment he caused the land to rise out of the global ocean. The sprouting implies a process by which the seed germinates and the sprout pushes through the soil to the surface to grow and leaf out ... Although there may have been an actual process take place, it took place at a much faster rate than it would thereafter in the normal order of things. (Barrick, forthcoming)*

All vegetation, including large trees, is said to have matured in the period of just a single day. More importantly, these processes seem to have had an *undeveloped* starting point. The vegetated world grew from seeds or at least from seedlings buried in soil, and then into fully mature, seed-bearing organisms. Danny Faulkner (2016) agrees:

> *It is clear that some aspects of instantaneous* **ex nihilo** *creation are within the creation account, such as the initial creation of matter in the beginning. But it also is clear that many aspects of the Creation Week involved processes. These processes are not to be confused with the supposed gradual, undirected, and naturalistic processes of evolution. Rather, the processes of Creation Week were rapid and directed by God to complete His purposes within one week.*

Faulkner (2016) has gone a step beyond that of Baumgardner (2000) in proposing some kind of highly accelerated processes for much of the creation. Faulkner gives his theory the name *dasha*, after the Hebrew word translated "to sprout" from Genesis 1:11 (ESV).

SFP theory and the constancy of relative rates of change

I completely agree with everything Westermann (1994), Baumgardner (2000), Faulkner (2016), and Barrick (forthcoming) have said above. Their commitment to supernatural creative processes during the Creation Week are, I think, essential to all that is conveyed within the term SFP. I would like to set the bar a little higher by incorporating one extra concept into SFP not discussed by these authors. It is not just about *processes*, or even *accelerated processes*. I propose the *constancy of relative rates of processes*. With this extra detail, I would like to develop the idea of SFP into a theory that incorporates all three of these characteristics. I will call this theory, supernatural formative processes theory or SFP theory.

Consider the rates at which two completely different kinds of modern plants grow. Relative to each other, one will likely grow at a faster rate. Now consider the same two plants growing on Day 3 of Creation Week. I propose that although they would grow at accelerated rates, their *relative rates* of growth would remain unchanged as the plants entered "normal" history.[2]

Observing Day 3 unfold would be kind of like looking at a time-lapse video. No *rogue* growth rates would catch your eye as you watched the plant realm spring up from seeds or seedlings. This may mean that fast-growing plants, like annuals, were brought into

---

2 Given the same or similar relative environmental conditions.

existence at the end of Day 3, while slow-growing plants, like the bristlecone pine, sprang up earlier in the Day.

Alternatively, it is *possible* that a time-lapse scenario for the organic realm involved the growth *and* death of some plants. In this scenario, all plant species were brought into existence at the same time early on Day 3. This would mean that annuals and other such transient plants would go through many generations, relative, for example, to the bristlecone pine.

Something similar seems to have occurred during Creation Week on continental shelves. Stromatolites are small, mound-like structures that typically accrete due to the work of photosynthetic cyanobacteria. Yet we find many horizons of stromatolites buried one atop the other in some of the earliest rocks on Earth (Stanley and Lucza 2015). These rocks are considered Creation Week rocks by many prominent YEC (Purdom and Snelling 2013). This means that during Creation Week, communities of cyanobacteria grew and died multiple times (as they were buried and replaced by new cyano-bacterial communities) all over the world. It also seems *possible* that multiple generations of stromatolites, as well as terrestrial forests, lived and died *for the purpose of secondarily* supplying the Earth with many of its life-sustaining resources, such as oxygen for its atmosphere and nutrients for its soils and sea floors (Coulson 2018). Snelling (2008) suggests similar divine purposes for the secondary origin of Earth's metallic reservoirs.

Of course, as the Christian philosopher Vern Poythress (2019) has correctly observed, time is relative. For example, if all rates everywhere in the universe were doubled, the participant would not notice this doubling unless there existed an unaltered rate that could serve as a frame of reference (Poythress 2019, p. 221). I propose that God has provided this frame of reference in the refrain that occurs at the end of each creative day: "And there was evening and there was morning." This would mean that *except* for the rhythm of night and day, all other rhythms were accelerated in a time-lapse fashion. Each *normal* cycle of day, evening, and morning demonstrates an incredibly powerful, universal frame of reference that ties the unusual events of Creation Week to ordinary human experience.

Time-lapsing of Creation Week can go only so far, however. Plants and cyanobacteria are not alive in the biblical sense, but animals are. Since death cannot precede the Fall (Genesis 1:30), a time-lapse scenario is out of the question for animals made on Days 5 and 6, including man. It would seem then, that for the inorganic realm, rates were accelerated billions of times, while for the plant realm the rates were accelerated only perhaps hundreds of times. The arrival of animals on Days 5 and 6 of Creation Week may have been the end catalyst for *exponentially decreasing rates* of change, allowing the animals to grow at rates that would soon be consistent for the Earth as it entered normal history.

Why keep relative rates of change constant?

Using processes, and even accelerated processes, seems to be required given God's desire to create a fully functional universe in six days. But why create the universe in a time-lapse fashion using a constancy of relative rates? I believe there are at least three theological reasons.

First, the supernatural differentiation of our planet occurred using a sequence of events that mirrors the same *naturalistic*[3] processes

3 I use the term naturalistic to emphasize God's providential rhythms of regularity in nature. God uses supernatural, non-rhythmic creative powers. But, according to Scripture, He also uses natural, rhythmic sustaining powers (natural laws). This is what I mean by naturalistic. Do not confuse naturalistic with naturalism. The latter term denies supernatural events and divine superintendence in nature.

operating today. This makes sense since the Earth was not designed to function supernaturally. It was designed to function *naturally*. God's *anticipating* these natural rhythms necessarily implies their being incorporated into the creative design in the first place.

Second, God gave Adam specific commands to subdue and have *dominion* over the creation: "Then God said, 'Let us make man in our image, after our likeness. And let them have dominion over the fish of the sea and over the birds of the heavens and over the livestock and over *all the earth*'" (Genesis 1:26 ESV; emphasis mine). How can humans do this if they do not understand how the creation functions? It makes sense, then, that the processes we see at work within the Earth today were the very same processes that God used to build it in the first place. This means that we can investigate God's creation using the processes that currently sustain it to find out how it ticks.

The third reason relates to the perspicuity of Scripture and God's use of analogies. Throughout the Bible God uses objects, historical narratives, and people in varied analogous ways for the purpose of conveying spiritual truth. It should not surprise us if God uses His present sustaining powers to teach us something about His past creative ones (see Poythress 2019).

There are also good scientific reasons for supporting an across-the-board constancy in relative rates of process. In the Pilbara region of Western Australia, scientists have found some of the "oldest" continental crust, which conventionally dates to about 3.5 billion years old (Stanley and Lucza 2015). At the edges of some of these ancient continental rocks are shallow-water stromatolitic reefs (Allwood et al. 2007). Stromatolites begin to form when sediment gets stuck in the sticky mucilaginous sheaths of photosynthetic cyanobacteria. Eventually, the sediment blocks out the sun, and in order for photosynthesis to continue, the entire community of microbes must move upward to populate a new surface. Over time, this accretion of sediment (actually sediment grains stuck in the old mucilaginous material of the microbes) turns into a stromatolite—a layered package of sediment that often, but not always, takes the form of columns and/or cones.[4] Most of those found in the geological record are anywhere from a few centimeters to several meters in height and superficially resemble some of the modern forms that are currently growing in Western Australia (Coulson et al. 2016; Coulson and Brand 2016).

Importantly, the Pilbara stromatolites did not accrete from particles of sand, silt, or clay. They accreted from particles of calcium carbonate, a specific kind of sediment that forms in *quiet* shallow *marine* environments. This means that these stromatolites most likely grew in a shallow marine, sub-tidal environment that was relatively free of sand, silt, and clay. We know this because geologists sampling pieces of these stromatolites determined that sand, silt, and clay were virtually absent (Allwood et al. 2007).

All of this is important from a YEC perspective because most YEC think these Pilbara rocks and associated stromatolites came into existence just after the continents emerged from the waters on Day 3 of Creation Week (Purdom and Snelling 2013; Clarey 2018). Here's the problem: Erosion associated with Day 3 regression of Earth's oceans would cause a staggering amount of sand, silt, and clay to be entrained and carried in the water column, perhaps for months or years. Most YEC think this erosion was rapid and catastrophic. Yet how did the stromatolites grow in a sub-tidal, *continental* environment that was almost completely free of sand,

---

4 Conical stromatolites are found only in Archean or Proterozoic rocks.

silt, and clay *on the same day* that trillions of tons of these sediments were eroded from the very same continents?

The growth of this stromatolitic reef during Creation Week can be accounted for *only* if the rate of continental erosion was accelerated, *along with* the rates of sediment deposition, *and* stromatolite growth.

This might mean that the Day 3 regression of Earth's oceans was not catastrophic at all. Something is geologically catastrophic when a *single rate*, say erosion due to a flood, is accelerated many times relative to other geologic rates—e.g., deposition of sediment, growth of plants, production of soil. Since *all the rates* were accelerated and remained constant with each other, an observer would witness the development of a genuine, shallow-water environment proceeding at break-neck, yet time-lapsed speed.

In reality, there were likely *dozens* of processes at work during the construction of this stromatolitic reef, all working at rates proportional to those operating today. A constancy of relative rates, or time-lapse scenario, is the only way to adequately account for the growth of this stromatolitic reef. Taken together, I believe YEC need to start thinking in terms of an initial *ex nihilo* creation followed by a time-lapse transformation of that creation.

Supernatural versus natural

I want to point out a common mistake that equates naturalistic laws with impersonal forces.[5] We must remember that it is God who sustains this universe using naturalistic laws that He Himself enforces. God is *personally* involved when it rains, when the crops grow, and when the sun shines. Naturalists have fooled themselves into thinking that these regularities are somehow impersonal.

---

5 See Vern Poythress (2019) for a good discussion on these concepts.

Yet Scripture declares that it is God who sends the rain (Mathew 5:45), it is God who controls the weather (Job 38:33 – 37), and it is God who upholds the very elements of the universe (Hebrews 1:3). God uses regular (naturalistic) laws and irregular (supernatural) laws to accomplish His will. Remarkably, many YEC treat natural, regular laws in an impersonal way. Yet God is just as much involved in causing a mountain to grow today as He was during Creation Week. God's using regular naturalistic mechanisms to form objects during the Creation Week in no way detracts from His power any more than it detracts from His power when sustaining the creation using the exact same mechanisms today.

Objections to SFP

It seems clear from the biblical evidence, as well as from the scientific evidence, that God chose to create the Earth by setting in motion a set of processes whereby the more "evolved" parts of the Earth (lighter crustal rocks) came from less "evolved" mantle rocks.

These concepts, although appealing to the more progressive YEC mindset, may seem almost heretical to those of us with a more conservative perspective. Some might argue that all these different rocks were *simultaneously* brought into existence at a single point in time (perhaps Day 2 or 3). In other words, an actual *sequence* of events that produced a light, granitic crust from heavier mantle rocks did not actually occur (contra Baumgardner 2000 and Faulkner 2016). Rather, all these different kinds of rocks sort of "just appeared" all at once (*ex nihilo*).

The creation of Adam is often used as an example. Clearly, Adam was created as a fully mature man from the dust of the ground (Genesis 2:7). Yet before Adam was created, he *was designed*. God did not just blink His eyes, wiggle His nose and bring a very complicated organism into existence without prior "thought." That God created Adam as a fully mature organism is, therefore, only *part* of the story. God *could* have created Adam as a baby, a young boy, or even an old man. The reason we automatically affirm this possibility is because we know that the creation of fully mature organisms was not God's *providential* intention. God designed humans to go through a natural process of maturation, starting life as a zygote, then developing into a fetus, progressing through various stages of infancy, childhood, pre-pubescence and pubescence before becoming a mature adult. This is God's *providential* design.

Yet where (or when) was this *developmental design* conceived? It certainly was not *after* God created Adam. Surely, God had designed all the different phases of human development *before* a single biological cell had sprung into existence. In other words, the entire life cycle of humans—from conception to death, was not only built into our genetic make-up, it was *designed in the mind of God before Adam was created.*

It turns out that in God's economy, *everything* is developing: humans, animals, bacteria, planets, stars, solar systems, and galaxies. The Earth *is* maturing (Chapter 5). Although God could have created Earth with a fully differentiated core, mantle, and crust, we must admit that somewhere back in time, even back into the very mind of God, the Earth's entire "life cycle," as with humans, had been meticulously planned out. God did not simply create the Earth as a static entity, he created it with a pre-planned commitment to sustain it, causing it to change, develop, and mature. It is one thing to create a static object, it is quite another to *design a dynamic system.* In order to accomplish the latter, the entire developmental process must be known *prior* to its unfolding.

Ex nihilo must, therefore, eventually lead back to process, even to "processes" that occurred in the recesses of God's mind. The core, mantle and crust may not have differentiated during Creation Week as I am positing here, but either way, God must have accomplished this feat *in His mind.* Don DeYoung, a prominent YEC, articulates these same ideas when he says, "*Processes* which in the mature creation viewpoint did not actually occur, such as a supernova explosion, remain *real events* in the mind of God" (DeYoung 2010; emphasis mine).

One objection relates to the *physical* interactions between the sun and the Earth during Creation Week. If rates of processes during Creation Week were highly accelerated, *yet still proportional to the same rates working today*, then surely the effects of the sun on the newly forming Earth would produce some rather strange phenomena. How, for example, would various seasonal plants grow when all plants all over the Earth are exposed to a *single* period of daylight and a *single* temperature representative of a *single* season (whatever that might have been)?

My response to these objections is simple and founded on the biblical examples of SFP found in Chapter 3. Almonds require variations in climate to produce a successful harvest. Yet the almonds that grew from Aaron's staff (Numbers 17:8) appear to have matured without the presence of these climatic variations. Remarkably, these almonds grew without needing sunlight, water, or even nutrients. These almonds matured, it would seem, *as though* the necessary natural conditions were present. Using other examples of SFP will secure the same conclusion. I propose that God caused the growth of a complete vegetated ecosystem *as though* all the natural mechanisms were present. The periods of day, evening, and morning served to fix the reader's temporal frame of reference,[6] *not to cause the growth of plants*. This explanation solves further objections related to the absence of the

---

6 Poythress (2019) disagrees with a calendar-day interpretation for Genesis 1. He dislikes terms such as "ordinary day," "literal day," and/or "24-hour day" because they inadvertently impose modern conceptions of time upon the text. As such, Poythress argues that YEC lack an *external* "standard of measurement" for the "length" of the Creation Week days (Poythress 2019, p. 256). Poythress' objection relates to the *length* of each Creation Week day as it correlates to the *length* of all subsequent "normal" days (like ours). If the text is not explicit on the *length* of each Creation Week day, then how can a YEC say that he is a "24-hour" creationist?

When YEC use these terms, however, they are not using them to suggest that God created the universe in exactly six 24-hour, 86,400-second days. They are correlating their own experience of time with a universal frame of reference that God has wisely incorporated into the creation account: "And there was evening and there was morning." This refrain occurs at the end of each of the six creative periods of daylight. This cycle of daylight, evening and morning *is* the "standard of measurement" that Poythress (2019, p. 256) thinks YEC lack.

This, of course, brings us to the reason *why* Poythress thinks that a temporal frame of reference does not appear in Genesis 1. Poythress interprets the refrain, "And there was evening and morning" in terms of a cycle, *not of time* but of *divine work* (daylight) and *divine rest* (evening and morning). The analogy is meant to correlate with a human period of work that occurs during the day and a human period of rest that occurs in the evening. As humans rest, so God rests.

Yet is this the best interpretation? Although God uses the Creation Week as an analogy for the human workweek (Exodus 20:9 – 11), I think it is a stretch to push the analogy any further. The text clearly says that God rested on the 7th Day (Genesis 2:2 – 3), but it nowhere indicates God's resting at any other time (i.e., at the end of each creative day). God could surely have included the words "and God rested" if, indeed, His intent was to draw an analogue with daily human rest. Even if the "God rested/humans should rest" metaphor were correct, there are no theological reasons to jettison the "ordinary" day meaning.

On the contrary, there is a very good reason for seeing these cycles in terms of ordinary days. Plants do not grow, mature, and bear fruit in a single day. The emergence of large tracts of land and oceans are usually experienced over millennia. A normal cycle of day, evening, and morning furnishes the reader with a powerful frame of reference that is universal across all cultures and times. This frame of reference helps constrain the temporal aspect of Creation Week despite the unfolding of astonishing processes.

sun on the first three days.[7] That the sun was not present during these days is inconsequential. What matters is the presence of the *temporal frame of reference*, day, evening, and morning. This frame of reference helps *constrain* the temporal aspect of Creation Week *despite* the unfolding of astonishing processes.

7 Excluding the Sun from the first three day/night cycles has a theological reason that is covered in Chapter 6.

Chapter 5

SFP and the Creation of the Earth

The *doing* of science

So why, in the last chapter, was there a rather lengthy discussion about *process* during Creation Week? The goal of this book is not simply to show that God used processes during Creation Week. The goal is to *apply* this concept to the real world, to build a scientific and theological synthesis that blends the supernatural of the Creation Week with the natural of the present world.

Consider the problem from the point of view of a Christian who happens to be a geophysicist. What is he to do with the *time* involved, at current rates, for all the processes that he knows are at work within the planet (e.g., fractionation, partial melting, and gravitational differentiation)? He is specially told that God created the Earth in six literal days, a clearly demarked time frame that extends over six 24-hour periods.

Now, the classic reply from the non-scientist is that God did it like that and we just need to believe Him. The problem with such a simplistic approach is that this Christian is a *geophysicist*. He knows that his fellow secular geophysicists are correctly able to do research based on the application of science to these natural processes, adding valuable information to the discipline of geophysics and at the same time contributing to many different fields of human endeavor. This is true for geophysicists, astronomers, mathematicians, biologists, geologists, etc. Although many of these fields do not expressly depend on processes that occurred in the past, some of them do and as such require a workable model from which to *do* science.

According to SFP theory, God used natural processes working at constant relative rates of change during Creation Week. Accordingly, our geophysicist should not be able to differentiate between a Creation Week rock and one formed subsequent to Creation Week. This would mean, for example, that the only difference between a secular interpretation for the origin of the Earth's crust and a YEC one is the *time* involved. Our Christian geophysicist, however, knows exactly how much time passed during the formation of the crust because he was *told*. It took anywhere from one to six literal, 24-hour-long days.

This temporal distinction, however, can be seen only with the eyes of *faith*: "By faith we understand that the universe was created by the Word of God" (Hebrews 11:3 ESV). The unbeliever will not accept this distinction. This is because his or her mind is closed to spiritual realities and a supernatural Creator. In the next section, we will see that keeping this distinction in place aids Christians who are also scientists in the *doing* of good, predictive science, while at the same time remaining faithful to a literal interpretation of the Bible.

Scientific anomalies

Given the importance of *process*, as well as a *constancy of relative rates* of process, our geophysicist who is also a Christian can successfully apply scientific methodology to the relationships that exist between the crust, mantle, and core. This, however, presents an interesting thought experiment: If rates for processes working during Creation Week were highly accelerated, *yet still proportional* to the same rates working today, then geological irregularities should be absent from Creation Week rocks.

What do I mean? Well, let's think about *gravitational differentiation*, a process that is related to the creation of continental crust but not integral to the process of partial melting (one of the major processes involved in generating continental crust). Summarizing the latter process from Chapter 4, partial melting works on the melting temperature of different minerals. Silica-rich minerals have a lower melting point. That means as the mantle heats up, quartz (a silica-rich mineral) will melt first. The heavier, metal-rich minerals will melt last or not at all if the temperature does not get high enough. What you end up with is a less dense, silica-rich fluid that eventually makes its way ("floats") to the Earth's surface and becomes crust.

Gravitational differentiation, although involved in the production of continental crust, is more involved in the formation of the Earth's dense core. Heavy elements, such as nickel and iron, typically do not migrate into the crust when they reach their melting points, but instead are drawn towards the Earth's center under the influence of gravity, hence the name, *gravitational differentiation* (separation: as in separating the core and the crust from the mantle).

Crudely, we can think of these separate processes (partial melting and gravitational differentiation) as sort of working together to *differentiate* the Earth into three basic parts: a core, a mantle, and a crust. If God, during the construction of the planet, chose not to keep the rates of these two processes *proportional* to the same two processes operating today, then you would end up with an anomaly that would not make sense scientifically.[1] This is because these two processes are based on principles of physics. Changing the rates means changing the physics. If you changed the rates of gravitational attraction, but not the rates at which minerals melt, you might end up with a layer or band of nickel and iron sitting *within* the mantle instead of residing at the Earth's center!

Since humans were meant to understand how gravity and partial melting work (Genesis 1:26 – 28), our Christian friend who is also a geophysicist would not be able to *do* science. The fact is, no one would be able to do science if such anomalies existed. In order for our geophysicist to do good, predictive science, the *current* rates at which partial melting and gravitational differentiation separate *today's* planet into core, mantle, and crust must have been working at the same rates *relative to each other* during Creation Week.

So, although these rates were accelerated, they were still interrelated as they are today. What does all this mean? It means that our Christian geophysicist can *do* science. It also means, however, that if our Christian geophysicist comes across an anomaly that is inconsistent with all the data as a *whole*, he should not automatically default to rates and processes foreign to those he is currently working with. This commonly occurs in YEC apologetics. An anomaly is found that is inconsistent with the majority of other phenomena and is hailed as "proof" that the Earth is 6,000 years old.

---

1 Assuming, of course, that the material started out as an undifferentiated mass. Importantly, Earth's differentiation into core, mantle, and crust using partial melting, fractionation, etc., is a *theory*. I think it is a good theory, but science is constantly changing, and such theories are likely to change with them. SFP theory does not validate these materialistic ideas. It simply allows Christians the freedom to *cautiously* use them.

Polonium halos[2]

One anomaly that accentuates the latter is polonium halos in primordial[3] rocks. Polonium halos are formed within minerals and their existence is due to nuclear decay. As the element polonium decays to lead, nuclear emissions (alpha particles) scar or etch the surrounding crystal structure of the mineral, producing a three-dimensional, colorized sphere (Snelling 2005). Think of shooting a bullet into ice: The bullet would cut through the ice, leaving a trail behind that you could see. Now imagine thousands of bullets fired in every possible direction. If the block of ice was large enough, all those bullet trails would create a three-dimensional sphere of damage in the ice.

What's interesting about polonium isotopes (there are three types) is that the rates at which they decay are incredibly short—on the order of seconds to minutes (Snelling 2005). Robert Gentry (1986) created a bit of a flurry back in the 1970s and 1980s when he proposed that these polonium halos/spheres were evidence for a supernatural, *ex nihilo* creation of primordial rocks. The rationale for this claim seemed solid enough. Since the rate of decay for any of the three isotopes of polonium is exceptionally short, the existence of the polonium halos (damaged rock) must require that the magma crystalized into rock within seconds or minutes after God created it. This is because large amounts of magma require hefty time frames to cool—typically thousands to millions of years. By that time, of course, all the polonium (each having a very short half-life) in the melt should have decayed away millions of years prior to the rock's crystallization.

2 Appendix B contains a simplified version of these processes and arguments for interested readers.

3 Rocks made during Creation Week.

When the rock finally crystalized, there should have been no polonium left to damage the rock. The halos, then, "prove" that the rock crystalized instantly. Although this hypothesis sustained some momentum in creation circles in the 1970s and 1980s, it has mostly fallen into obscurity with the advent of a simple solution.[4] All three polonium isotopes are produced in the decay chain of uranium 238 (Snelling 2005). It seems most likely that the gas radon, the element that presents itself in the uranium 238 decay chain prior to polonium, migrated away from the original site of uranium decay in tiny flaws found in the mineral biotite. Once the radon (a gas) decayed to polonium (a solid), the rest of the decay chain unfolded at this new site *separate* from the uranium halo—hence a polonium halo.

Earth's decaying magnetic field[5]

Another anomaly is the decay of the Earth's magnetic field. Unlike the example above, this anomaly is commonly used in YEC communities. Some background information will be provided for the purpose of highlighting the problem.

The Dynamo Effect (a secular theory) essentially holds that Earth's magnetic field reverses its polarity every few hundred thousand years. This means that our current magnetic north pole[6] "flips" to become a magnetic south pole, and the magnetic south pole flips to become a magnetic north pole. Think of a very large

---

4 Gentry's research is still useful in creationist apologetics since he also discusses polonium halos in other objects, like wood, for example. These findings have other important implications. Snelling (2005) has also added valuable new research on the importance of polonium halos in metamorphic rocks.

5 I suggest that Appendix B, which contains a simplified version of these processes, be read first for those with no background in this subject.

6 Actually, a "North-seeking" pole.

bar magnet located between the *geographic* north and south poles that literally flips ends. The overall effect over millions and billions of years would be an oscillatory curve that could be modeled as a sinusoidal wave (Figure 1A). There exists some good evidence to support the validity of this model—for example, magnetic striping of the Earth's ocean floors and the interesting observation that the sun reverses its poles every 11 years (Stromberg 2013).

Most YEC agree that the Earth is reversing its polarities but *limit* these reversals to events associated with Noah's Flood (Figure 1C). In this model, continental plates were rapidly subducted into the mantle during the Flood of Noah causing instabilities within the core. These instabilities were responsible for rapid pole reversals but were eventually damped out over several centuries. This means that the Earth was *originally* created with a *single* polarity (Figure 1B) but that the direction of this polarity was *temporarily* altered during the Flood of Noah (this is why figure 1C has "squiggles" that interrupt the nice curve shown in figure 1B).

Given this background, we are now able to address the problem. As it turns out, the strength of the Earth's magnetic field is *decreasing* over time. Extrapolating this rate of decay backwards in time furnishes an "age" for the Earth somewhere in the vicinity of about 10,000 years (Humphries 1984)—a nice fit from a YEC perspective. Moreover, given a *single* polarity direction (North), a backwards extrapolation would *increase* the field strength to thresholds at which *life* was not sustainable, thus making Darwinian evolution impossible (Figure 1D). These claims by YEC are only valid, however, *if* the Earth has experienced a *single* polarity (Figure 1B). It is true, there were some temporary pole reversals during the Flood (Figure 1C), but these reversals in no way affected the *general* decay and/or direction of the Earth's northward-pointing magnetic field.

Secular scientists agree that the Earth's magnetic field is decaying rather dramatically but insist that this is normal. As stated above, they believe that Earth's magnetic field has reversed its polarity hundreds of times in the past (the Dynamo Effect), and that the existing decay of the field merely represents a *single reversal event*. Once it "flips," the magnetic field will *increase in strength* before the cycle repeats itself (Figure 1A).

Many YEC, however, maintain that no mechanism exists capable of *regenerating* the Earth's magnetic field. If this is true, then it would be impossible for the Earth's magnetic poles to reverse, *even once*. This leaves the secular scientific community without an explanation for the reversals inscribed on the seafloor (see Appendix B). The overall result is quite advantageous to the YEC model, since it provides both a mechanism for reversals (rapidly subducting plates) and a definitive time frame for the field's (and thus the Earth's) existence—about 10,000 years. Secular scientists have recognized the problem and although they have some promising models (O'Rourke and Stevenson 2016), must admit that they are only at the hypothesis stage.

Applying SFP theory

Let us consider the polonium halos. In order for our geophysicist to acknowledge the validity of such a hypothesis—that these halos "prove" the instantaneous crystallization of Creation Week rocks—he must also dismiss any other data that exists *within* these rocks. This is because the rate at which the rocks cooled proceeded *independently* from the rate at which gravity was pulling on the minerals and elements within the rocks. These rates were also working *independently* of the rates at which isotopes decay and heat is dissipated. This then sets up a domino effect that cascades into every other area of geophysics.

Essentially, our Christian geophysicist cannot *do* science well. This example is diametrically opposed to SFP theory.

Now let us consider the decay of the Earth's magnetic field. Summarizing from above, the Earth has never had a *self-generating* mechanism for reversing its magnetic poles (Humphreys 1984). This means that the present decay of the Earth's magnetic field is not related to the Dynamo Effect (continuous oscillations for billions of years). Rather, the Earth has *always* had a *single* polarity from Creation Week onwards, and the present decay associated with this field began at Earth's inception during Creation Week (Figure 1B). Extrapolating this rate of decay backwards furnishes an "age" for the Earth of about 10,000 years.

Given these data, what is our geophysicist to think? What should he say to his Christian friends and colleagues? Although this rate of decay affirms a very young Earth, he is also aware of other rates of change that if extrapolated backwards seem to affirm the antiquity of the Earth (e.g., partial melting, fractionation, gravitational differentiation, radioisotope decay). Should our geophysicist use the magnetic field decay rate to *affirm* the youth of the Earth given these contradictory data?

Let's investigate things a little more. Using SFP theory, our geophysicist knows that the Earth's magnetic field was established during Creation Week and that the generation of this field was incumbent upon many other processes directly related to the differentiation (separation) of the Earth's internal structure—processes like partial melting, fractionation, gravitational differentiation, nuclear isotope decay, and multiple others required to produce a hot, liquid core.

Our geophysicist also acknowledges that all these different rates, *including* the rates responsible for the initial generation of the magnetic field, must all have been constant *relative to each other*. So, if partial melting and fractionation were accelerated, so too was the rate at which the initial magnetic field was generated.

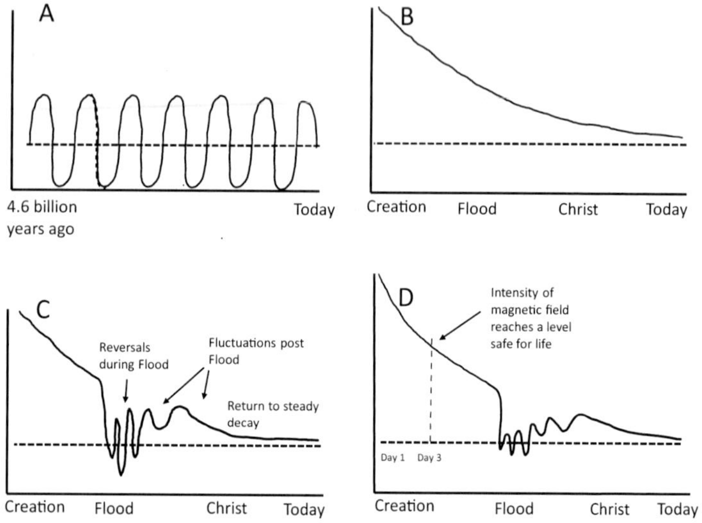

Figure 1. Understanding magnetic reversals from a creationist perspective. The Y axis is intensity of field, the X axis is time. The dotted line represents a magnetic field strength of zero. When the field strength reaches zero, it will gain strength in the negative direction. A. The secular model contends that reversals have been going on for 4.6 billion years (there have been hundreds). A single, North pole direction is highlighted. As the field weakens, it passes through the dotted line and begins to gain strength once again. At the bottom of the curve, the pole flips, and then begins to lose strength on its way back up to the dotted line. B. Humphrey's model suggests a single North pole direction from Creation Week that slowly decays. C. Humphreys adapts his model to explain the "fossil" reversals recorded on the ocean floor. D. Humphreys' model can still work since the rate at which the field decays is the same today, relative to other rates, as it was during Creation Week (note, however, that since rates were highly accelerated, the slope of the line prior to Day 6 would be almost vertical). God chose to bring life into the new creation only when the magnetic field strength was suitable for life to exist. Adapted from Humphreys 1989.

Since this polarity has a single direction (North), however, extrapolating it backwards into Creation Week would seem to suggest an infinite field strength with a curve that would eventually go vertical! (Figure 1D). There is also the problem of newly created organisms.

Extrapolating a single magnetic polarity backwards would increase its strength to thresholds at which life was not sustainable during Creation Week. How could the Day-3 biosphere tolerate these dangerous extremes?

As it turns out, such a scenario would not generate an *infinitely* powerful magnetic field because magnetism is a property of atoms. The core can contain only a *finite* number of atoms and so can generate only a *finite* magnetic field. Although this field was not infinite, however, it would still have been deadly to the Creation Week biosphere. Our geophysicist could propose that during the Earth's development, there came a point within Creation Week at which the magnetic field's strength decreased to levels *fit for life* (Figure 1D). This would correspond to Day 3 of Creation Week.

These observations suggest that Humphreys' hypothesis *could* work using SFP theory.

Extrapolating a single polarity backwards in time will not produce an infinite field, and although it might have been dangerously high prior to Day 3 of Creation Week, no damage to living organisms would have occurred because no living organisms existed.

We must, however, keep in mind that using this example to "prove" that the Earth is young opens up a Pandora's Box of other problems. What is the YEC to do with the *other* rates? Extrapolated backwards, rates of fractionation, partial melting, gravitational differentiation, and radioisotope decay equally *contradict* the rate at which the magnetic field is decaying. Most YEC are aware of these other rates. Baumgardner (2000, p. 87), for example, says this about radioisotope decay: "I propose that on the order of 4.5 billion years' worth of nuclear decay at present rates unfolded rapidly in the earth before the end of Day 3." Is it appropriate, therefore, for YEC to emphasize the rate at which the Earth's magnetic field decays without addressing many other rates that "prove" the Earth is millions or billions of years old?

As noted above, there is a robustness to the current dynamo model. Magnetic seafloor striping is consistent with a sinusoidal oscillation for pole reversals. Taken in conjunction with the sun's pole reversals and the reality of a hot, liquid outer core conducive to convection currents, self-generated magnetic pole reversals on Earth seem *possible.*

Given all the above, I *personally* believe that our geophysicist should *not* use this example to affirm the youth of the Earth. It is an interesting anomaly, but since a multitude of other rates equally contradict these data, the result is equivocal. For those that insist on using this example, I suggest fair play and a Christian transparency that doesn't stack the deck.

Does the Earth have a birthday?

At this juncture, I think it pertinent to address some possible misconceptions. I, of course, am not suggesting that the Earth is 4.6 billion years old. Neither am I suggesting that the "apparent age" of the Earth is 4.6 billion years. Secular scientists have arrived at a 4.6 billion-year-old age for the Earth because in their minds, radioisotope decay models seem to suggest that, indeed, this is the correct number.

Is it possible, however, that since God used interdependent "synced" processes, that some kind of quantifiable number, even if apparent, can be applied to the formation of our planet? Note, I am not saying that this *number* has anything to do with comparison—like looking at another planet and saying that that planet looks 4 billion years old so ours must be a smidge older. Neither am I saying that God made the Earth *look* old. This of course is nonsense. God formed the Earth using process, and process is always interpreted by humans as the passage of time.

That God intended us to recognize Adam's "apparent age" can be derived from the text. Although Adam was only a few hours old at the end of Day 6, he was *not* created as an infant or an old man. He was created as a sexually mature adult. We do not need to use *years*, therefore, to describe this apparent age. We could simply use three descriptors: immature, mature, or elderly. Is it possible that our Earth was likewise created to have an "apparent age" that we could describe as "mature"? What about describing our sun as a "mature" star? In cosmological lingo we might call it a main-sequence star, meaning that it is in the most stable or "mature" phase of its existence. Immature, mature, and old are not subjective abstractions; they are physical realities that are intrinsic to the organic as well as the inorganic realm. I could, therefore, use these three stages of maturity to describe planets, stars, solar systems, and/or galaxies. Given this, I think it would be acceptable to use a numerical scale *in years* to help us relate these three different stages, much like we use a numerical scale in *degrees* to relate a liquid's freezing and boiling points.

What is at stake when accepting or rejecting an apparent age for the Earth? Let's break it down a little:

1. If there is *no* apparent age for the Earth, this would mean that various processes *currently* at work within the Earth cannot be extrapolated backwards in time for the purpose of determining a "beginning." In this scenario, some of the Earth's *internal* processes can be extrapolated backwards millions or billions of years (partial melting, fractionation, gravitational differentiation, radioisotope decay), others can be extrapolated backwards hundreds of thousands or even thousands of

years (the decay of the Earth's magnetic field), while others cannot even be correlated at all since they were clearly supernatural acts (polonium halos).[7] This would mean that God used other, as yet unknown, mechanisms in conjunction with known processes when forming the planet.

2. If there *is* an "apparent age" of the Earth, this would mean that various processes at work within the Earth *can* be extrapolated backwards in time for the purpose of determining a "beginning." In this scenario, *all* the Earth's internal processes can be extrapolated backwards into Creation Week because, although these processes were accelerated, they were operating at the same rates, *relative to each other*, as they do in the present (SFP theory).

Choosing the first option has the benefit of underscoring a few processes for the purpose of "proving" that the Earth's *internal* structure was supernaturally created about 6,000 to 10,000 years ago. This option is perfectly valid since there are no theological and/

---

7 It is possible that the systems at work within and upon the Earth worked differently before the Fall. Some YEC have suggested that the pre-Fall world was in a constant state of *restoration*. These restorative processes were necessary given the opposing processes related to the Second Law of Thermodynamics. This would mean that all processes at work in the pre-Fall world were perfectly cyclical, having neither "beginning" nor "end." As an example, consider a stream flowing from the Garden of Eden. As the stream necessarily erodes sand and mud particles, transporting them to the sea, some process on the seabed would have to constrain these particles, transporting them back to the source at the river's head. This position would suggest that the Fall "broke" many of these cycles. Such restorative forces would necessarily make it impossible to precisely extrapolate rates of change backwards in time, since the extrapolation would be taking the investigator back to the Fall, *not* creation. See, however, my thoughts below.

or philosophical contradictions. God can create the universe in any fashion He so pleases.

Present day processes, for example fractionation and gravitational differentiation, could have had their rates sped up or slowed down, *relative to each other*, during Creation Week. This would be a bit like speeding up the rate at which mantle rocks melt (producing a light magma) without speeding up the rate at which heavy elements sink into the core by gravity. Evidence for such disparate rate changes would, of course, be evident to the modern geophysicist in the form of unexplainable "anomalies."

Those who choose this first option, however, must, by virtue of this choice, also agree that supernatural mechanisms (supernaturally speeding up one rate without speeding up the other rate), unexplainable using scientific methodology, were also at play during Creation Week. This means scientists cannot rely on *current natural* rates for the purpose of understanding anything that was made during Creation Week.

Choosing option two eliminates the advantage of "proving" that the Earth's *internal* parts are 6,000 to 10,000 years old, but does provide our geophysicist with a sound, scientific methodology. This is because he is committed to a set of interdependent processes that proportionally bridge Creation Week with the time/historical context of Day 7 onward.

This would mean our geophysicist *should not expect* anomalies. If an anomaly is found (e.g., decay of the Earth's magnetic field and/or polonium halos), then our friend should not automatically assume that this evidence "proves" the Earth is 6,000 to 10,000 years old.

Remember, I am not saying our scientist must adopt a 4.6-billion-year "apparent age" for the Earth. Instead, our geophysicist considers *all* the data for the purpose of approaching the *correct* "apparent age," which may be far less than 4.6 billion years.

There are many other methods for evaluating the age of the Earth. YEC Bodie Hodge (2007) has compiled a list of uniformitarian processes that, if extrapolated backwards, furnish differing ages for the Earth. These extrapolated "ages" range anywhere from a few thousand years to about a half billion years.[8]

If an "apparent age" for the Earth exists, then YEC should be seeking to correlate *all* the data for the purpose of finding out what that "apparent age" might be.

The YEC must remain consistent whether he or she chooses option one or two. Using science to "prove" that the Earth's *internal* structure is 6,000 to 10,000 years old based on the backward extrapolation of a few processes while neglecting other extrapolated processes (partial melting, fractionation, gravitational differentiation, radioisotope decay) is not only contradictory, but also unhelpful to our geophysicist seeking to *do* science.

The Fall and God's Curse on creation

What about God's Curse? If we assume that God created the Earth *in anticipation* of the natural laws that would soon sustain it, what changed as a result of the Fall and God's Curse upon creation? After all, it would seem as though processes currently at work within the Earth to sustain it closely resemble those processes that God used to create it in the first place (Chapter 4; Baumgardner 2000). How does God's Curse on creation fit into a nice Creation Week/

8 In Appendix A, I have also compiled a list of "ages" for the Earth *and* the cosmos based on the extrapolation of current rates of change.

pre-Fall/post-Fall transition? What happened during that very brief *pre-Fall* period?

First, it will be important to summarize what we know about the Fall and God's subsequent Curse.

We know from Genesis 3 that the biological realm was in some way altered as a consequence of the Curse: "cursed are you [to the snake] *above all livestock* and *above all beasts of the field*" (Genesis 3:14 ESV; emphasis mine). Notice that the livestock and the beasts of the field were cursed along with the snake, to a lesser extent to be sure, but nevertheless cursed. In Genesis 3:17 – 18 (ESV) we are told that the ground is cursed: "cursed is the ground because of you; in pain you shall eat of it all the days of your life; thorns and thistles it shall bring forth for you; and you shall eat the plants of the field." The Curse on the ground affected the relationship between the soil, the plant realm and man, so much so that plants developed defensive mechanisms, "thorns and thistles." That the entire creation was in some way affected by the Curse is clear from Romans 8:19 – 23 (ESV):

> *For the creation waits with eager longing for the revealing of the sons of God. For the creation was subjected to futility, not willingly, but because of him who subjected it, in hope that the creation itself will be set free from its bondage to corruption and obtain the freedom of the glory of the children of God. For we know that the* ***whole creation has been groaning*** *together in the pains of childbirth until now. And not only the creation, but we ourselves, who have the firstfruits of the Spirit, groan inwardly as we wait eagerly for adoption as sons, the redemption of our bodies (emphasis mine).*

Paul more than likely speaks of "futility" in terms of creation's inability to attain the ends for which it was made—a clear reference to Genesis 3 (Moo 1996). In verse 23, Paul distinguishes between "the creation" and fallen humans ("we ourselves"), so Paul must be referencing the *rest* of the organic realm as well as the entire inorganic realm.

That the animal kingdom has been affected by the Curse is evident from Genesis 3:14 (see above). These effects would include carnivory (Genesis 1:30), disease, and suffering in general. But what about the inanimate realm? In what sense is it "groaning"? Any cursory overview of ancient Greco-Roman history will uncover a regular occurrence of natural disasters, specifically related to earthquakes, volcanic eruptions, tsunamis and floods. Paul can say that these Christians "know" these things because they knew about or experienced these adversities.

Apart from the physical aspects of the Curse, there was also a spiritual component. Clearly, prior to the Fall, there was an intimacy that the physical realm experienced with its Creator. That intimacy was broken at the Fall and remains broken to this day. The consequences of this spiritual separation are described by the prophet Jeremiah in Jeremiah 17:9 (ESV): "The heart is deceitful above all things, and desperately sick; who can understand it?" And Paul in Romans 3:11 (ESV): "None is righteous, no, not one; no one understands; no one seeks for God." Clearly, fallen humans seek to glorify themselves and fulfill their own desires, abandoning the Creator God altogether.

Some well-meaning Christians have proposed that God cursed the universe by introducing it to the Second Law of Thermodynamics. But as Lisle (2009) wisely points out, the Second Law of Thermodynamics is beneficial in many ways. Without it, we could not receive heat from the sun, digest our food, or even walk! Lisle (2009) postulates that, "although the second law was in full effect, another compensating restorative process may have prevented any net decay of the universe." This restorative process ceased as a result of God's Curse, allowing things to deteriorate.

Exactly what this restorative process was and how it worked, however, is a matter of speculation. This speculation extends to large-scale processes that are currently in operation today. Exactly how, for example, would plate tectonics be affected by this restorative process? We know from the geological record that plate tectonics, although operating differently during Creation Week, nevertheless was in operation.

We must entertain the possibility that no such restorative process was at work at all. One of the main objections to such a notion relates to the eternality of the Earth. If, for example, the Fall had not occurred, the unfallen human race would necessarily have lived forever. Yet that would require an eternal home upon which to dwell. This presents a most interesting theological problem. Since the Second Law of Thermodynamics acts to bring disorder to an orderly universe, then that would suggest that the Earth, itself subject to disorder, will inevitably lose its internal heat to the universe, leaving it cold and barren. The same fate awaits our sun. Without a restorative process, the Earth and the universe will ultimately die.

Would the Earth, however, have persisted forever? Two biblical texts are used to justify this position. Psalm 78:69 (ESV) says: "He built his sanctuary like the high heavens, like the earth, which he has founded forever," and in Ecclesiastes 1:4 (ESV) we read: "A generation goes, and a generation comes, but the earth remains forever." The context of Psalm 78:69, however, is clearly one of *comparison*—the sanctuary, unlike its predecessor, the tabernacle, will no longer be moved around. It will remain stable. What better way of *poetically* illustrating this stability than with a comparison to the Earth—itself extremely stable, having a *sense of apparent* endlessness. Undoubtedly, the psalmist is *not* making a scientific statement of fact about the longevity of the Earth. The Ecclesiastes passage fits into the same category, making a comparison rather than defining a scientific statement. The passage contrasts the ephemeral nature of man with that of the Earth. In comparison to the *apparent* unfading Earth, man is but a vapor.

It is also important to realize that the Hebrew word "forever" (*'olam*), can just mean a very long time. Context must determine its meaning. Consider these verses from 1 Samuel 1:21 – 22 (ESV) compared with verse 28:

> *The man Elkanah and all his house went up to offer to the Lord the yearly sacrifice and to pay his vow. But Hannah did not go up, for she said to her husband, "As soon as the child is weaned, I will bring him, so that he may appear in the presence of the Lord and dwell there* ***forever ['olam]****" (emphasis mine).*

In verse 28, however, Hannah says: "Therefore I have lent him to the Lord. *As long as he lives ['olam]*, he is lent to the Lord" (emphasis mine).

So, if the Earth was not designed to last for eternity, at least in its present state, what do we do with a perpetual race of unfallen humans who live forever? One school of thought contends that humanity was never meant to live in its "good," yet far from perfect, state forever. In an unfallen world, this school contends that given an unspecified amount of time ordained by God, Adam and his offspring would, as with redeemed humans today, be changed from their unfallen, *yet corruptible*, state to the *incorruptible* state spoken of by Paul in 1 Corinthians 15. Saint Augustine (354 – 430) puts it this way:

> *Still, although it was by reason of his body that he was dust, and although he bare about the natural body in which he was created, he would, if he had not sinned, have been changed into a spiritual body, and would have passed into the incorruptible state, which is promised to the faithful and the saints, without the peril of death.*

This position has much to commend it. First, it alleviates the possibility of unfallen humans having to endure persistent temptation for all eternity. In 1 Corinthians 15:52 – 53 (KJV), Paul clearly teaches us that resurrected humans will not be able to sin:

> *In a moment, in the twinkling of an eye, at the last trump: for the trumpet shall sound, and the dead shall be raised incorruptible [unable to be corrupted], and we shall be changed. For this corruptible [able to be corrupted] must put on incorruption, and this mortal must put on immortality.*

This contrasts sharply with Adam. Even in his unfallen state, Adam was clearly *corruptible* (able to be corrupted). Although Adam could have lived forever, it appears as though God's ultimate intention included his transition to the incorruptible state after an undefined period of "testing." God may have subjected the angelic realm to a similar "test" resulting in the confirmation of fallen and unfallen angels to their respective states of choice. This is likely why the Bible describes only a *single* historical act of disobedience on the part of the angelic realm.

Given Augustine's position, God could have created the universe to last for a very long period of time, but not for eternity. Once mankind had reached its collective goal of obedience, God would transform both the organic and the inorganic realm, making both fit for the eternal state. This means that the Curse need not have necessarily altered the rest of the *physical* universe.

Of course, the text clearly tells us that the ground was cursed by God, but in what way? Apart from the biological facets outlined above, I propose that some aspects of God's Curse related to the onset of plate tectonics. This would make sense since many natural disasters, including earthquakes, volcanic eruptions, tsunamis, and flooding, are directly correlated with moving, subducting, and diverging crustal plates. I contend that although God used moving plates to construct the original crust during Creation Week, this movement ceased in preparation for the creation of the biosphere, including man, only to be reactivated soon afterwards as a result of God's Curse.[9]

It is possible that this tectonic reactivation was the catalyst God used to bring about the global Flood of Noah. This hypothesis allows for continuity between Creation Week processes and those that have been at work within the Earth since the end of Creation Week, excluding the geologically insignificant time represented by Adam and Eve's unfallen state.

---

9 Rigid plates would not affect the atmosphere and/or the livability of the planet, at least not in the short term.

Chapter 6

SFP and the Creation of the Universe

The conceptual universe

For the most part, SFP theory as discussed in Chapter 5 is straightforward. Things become a lot more theoretical when unpacking all that is involved in a similar approach for the universe. The reason for this is the order of creative events as they unfold in Genesis 1.

The text plainly teaches that the Earth was created first on Day 1, while the sun, moon, and stars were created at a later stage on Day 4. This order of events is, of course, in conflict with the current secular view that asserts the stars came first, then the sun, then the Earth. If the biblical order of creation followed the secular account, then SFP theory as applied to the formation of the Earth could also be applied to the universe. Since, however, the Earth *precedes* the rest of the universe in the creation account, one must take another step backwards and enter *the very mind of God*.

As such, the major premise behind all that follows is dependent on what was going on in the mind of God *before* Creation Week. This should not seem strange since we know from Scripture that many important ideas were formulated in God's mind before He brought forth a single created thing. For example, we are told that God made *promises*: "In hope of eternal life which God, who cannot lie, *promised before the ages began*" (Titus 1:2 ESV; emphasis mine). He *chose* the elect: "Just as He chose us in Him before the foundation of the world" (Ephesians 1:4 ESV). And He *loved* Jesus: "For you loved me before the foundation of the world" (John 17:24 ESV).

Even the functioning of Adam's biology, right down to the intricacies of the living cell, must have been planned out to the finest detail before Adam was created. We know this because the whole plan of salvation was set in motion before a single atom had sprung into existence. Yet the primary aspect of that plan was incumbent on the second person of the Trinity *taking on flesh*. God could not have planned all the redemptive aspects related to salvation without also planning the biological ones. It should not surprise us, then, that God had thought about His creation *before* He brought it into the time/historical context of Creation Week.

God is a master artist. All artists have an idea of what they are about to create before they put paint to canvas. An engineer does the same thing, conceptualizing in his mind the workings of the machine he hopes to bring to life *prior* to even sketching the plans. The point is that a significant amount of thought and planning goes into any creative process, and we should expect no less from God.

That divine archetypes exist in the mind of God, springing forth at His spoken Word to become planets, stars, humans, atoms, galaxies, plants, energy, etc., has also been discussed by various theologians. In Hebrews 11:3 (ESV) we read, "By faith we understand that the universe was created by the word of God, so that what is seen was not made out of *things that are visible*" (emphasis mine). Franz Delitzsch's (1874 – 1876, p. 215 – 223) *Commentary on the Epistle to the Hebrews* says this in relation to the latter half of the verse:

> *These ... [things not visible] are the divine ideas from which the visible universe sprang into being, and which were drawn from their divine seclusion in the mind of the Creator, by means of the creative word, into the region of mundane phenomenal reality.*

Philip Schaff's (1879 – 1890) *Popular Commentary on the New Testament* says this:

> *... or it may describe the purpose of the Creator, "in order that" what is seen might be understood to have come from what does not appear—viz., from the Divine mind and plan.*

There is also the *nature* of God's thoughts. Are God's thoughts, unlike the thoughts of humans, in some sense *real*? What does it mean for something to "exist" before anything ever did?

Humans think in terms of *existence* for all things pertaining to a material universe. Without really thinking about it, that which preceded this material universe is thought of as the opposite of existence, as though nothing at all existed prior to creation. Yet this is not strictly true. God and His thoughts existed prior to creation. This means that although existence, as perceived by humans, differs from existence as perceived by God, we must still be willing to say that, in some unfathomable way, reality and existence do indeed transcend the time/historical context of the created universe.

As an example of this, consider the man-made tabernacle that was patterned on a pre-existing archetype in heaven. The writer to the Hebrews (9:23 ESV) says, "Thus it was necessary for the copies of the heavenly things to be purified with these rites, but the heavenly things themselves with better sacrifices than these." Regarding this verse, Adam Clarke (1810 – 1826) says:

> *These are called patterns ...* ***exemplars****, earthly things, which were the representatives of heavenly things. And there is no doubt that everything in the tabernacle, its parts, divisions, utensils, ministry, etc., as appointed by God, were representations of celestial matters; but how far and in what way we cannot now see.*

Clearly some kind of spiritual tabernacle existed in heaven before the earthly tabernacle upon which it was patterned was ever even built. Incidentally, this verse also predicates a fundamentally divine methodology that corresponds to His creative design—God uses *patterns* much like humans do when constructing things.

Given the preceding caveats, I would like to propose that a fully functional, conceptual universe, already "existed" *in the mind of God* prior to Genesis 1:1. The Bible hints at this when it applies the anthropomorphic "God said" throughout the creative account. Before God "speaks," He has a *mind* with which He *thinks*.

It will be important to affirm that the idea of a conceptual universe is *hypothetical*. I am not saying, "this is how God did it," nor can I give chapter and verse necessarily supporting this conjecture. I am merely presenting a theoretical idea for what seems *plausible* given all that has been said above.

Constructing the conceptual universe

Many creationists agree that the universe is expanding, and galaxies are unfolding (Lisle 2011; Faulkner 2018a). Paul Garner (2009) says: "The limits of this extrapolation [the universe back to a singularity] can obviously be questioned, but the conclusion that the universe has undergone some degree of expansion from a smaller size seems robust."

The Bible itself seems to indicate an expanding universe: "It is he who sits above the circle of the earth, and its inhabitants are like grasshoppers; who stretches out the heavens like a curtain, and spreads them like a tent to dwell in" (Isaiah 40:22 ESV). As Lisle (2011) points out, this expansion has most likely been occurring since Creation Week.

I suggest that this expansion was first set in motion in the mind of God *before* the universe was placed into the space/time historical context of Day 4. This, of course, presents some interesting physics-related problems related to space, time, and motion. Given Einstein's general theory of relativity, many modern scientists define time in terms of *motion*. Since time and space are intricately woven together, you simply cannot have motion without the passing of time. Things, however, are not that simple. Theoretical physicists and philosophers disagree about the nature of time, as can be attested by a plethora of philosophical arguments that populate multiple websites and blogs. Until someone distills a sound, scientifically established definition for *time*, an expanding conceptual universe existing in the mind of God is not that philosophically far-fetched.

So, with what was this conceptual universe constructed? It almost certainly was not constructed from matter, at least not in the way we presently define it. Was it some unknown *spiritual* substance? Or perhaps this universe was just constructed in the same way humans construct things in their minds, where there is an image, but the thing formed is not tangible? The answer is really unimportant. What is important is that God, unlike humans, can "create" anything He likes in His mind, and that this "creation" is, at least in some sense, *real.*

It will be important to distinguish this conceptual universe from incorrect metaphysical ideas associated with pre-existing matter. Plato, for example, clearly believed in the eternality of matter (Berkhof 1938). Similarly, some may suspect an affinity with Plato's cosmology that took the world of Becoming (the physical universe) to be a copy of the world of Being (the eternal *Ideas*, or *Forms* that represent what is "real") (Bradie and Duncan 1997).

I am in no way endorsing these unbiblical metaphysical ideas. Plato believed that the present, physical universe is an imperfect copy of the impersonal "eternal *Ideal.*" God is not impersonal, and He does not consist of eternal matter. The conceptual universe I am describing was a *finite* product of God's thoughts in no way dissimilar to Adam's conceptual biology. Notwithstanding this, it would be a mistake to equate God's thoughts with human thoughts. Our thoughts are never *real.* That is because they are *dependent* upon that which is real, the created order, which itself exists independently from our thoughts. When humans die, their thoughts perish, but the created order does not. God's thoughts are not dependent upon anything. They exist independently from any other reference and so must themselves be representative of that which is in some sense *real.* In other words, God does not "look" at an object and frame an image in His mind characteristic of that object. He frames the image *first*, and then, as His will permits, "speaks" the image into *our* existence.

Which conceptual universe?

Was this conceptual universe static or dynamic? Many secular scientists still believe in a steady state universe.[1] Given this, which conceptual cosmology am I advocating—a Big Bang universe, a steady state universe, or other less popular cosmologies? The answer is *none* of them.

There are a multitude of problems associated with all cosmological models. The reason, of course, stems from a lack of experience with evolving planets, stars, and galaxies. That scientists cannot see or even detect about 96% of the known universe only compounds the problem (Moskowitz 2011).

Extrapolating cosmological rates of change backwards into the distant past will, therefore, be fraught with difficulties regardless of one's religious or scientific presuppositions. Given this, I think it extremely presumptuous of modern humans to think that the current leading cosmological model—the Big Bang—must be *the* metaphysical truth upon which all other observations rest.

Those who think the Big Bang to be *intrinsically* true are, as Poythress (2019) observes, fooled into thinking that ours is the correct vantage point from which to assess what is actual or real. So easily do we forget the myriad of past scientific constructs that at those times were viewed as fundamentally accurate—theories like Ptolemaic astronomy, Newtonian mechanics, classical thermodynamics, wave optics, humoral theory of medicine, phlogiston theory of chemistry, spontaneous generation, the theory of mountain building prior to the 1960s, and many more (Moreland 1989; Kuhn 1996).

---

1 It is important to note that all steady state cosmologies, even those held by creationists, must still make room for an expanding universe. See Faulkner (2018a).

Given these caveats, I propose that the conceptual universe God formed in His mind looked something like the one we presently observe. We can *confidently* use this as a starting point, perhaps winding things back a little to allow for colliding galaxies and exploding supernovae (see below). It would be unwise, however, for the Christian cosmologist to jettison current YEC models in favor of a conceptual Big Bang cosmology. An exploding conceptual singularity *may* be the metaphysical foundation upon which reality rests, but it also *may not* be. As stated above, there are simply too many holes in our cosmological knowledge to warrant such extrapolations.

Concisely, without a conceptual universe, a Christian cosmologist committed to a Day 4 creation of the sun cannot consider a Big Bang cosmology. Given the *possibility* of a conceptual universe, however, and the Christian cosmologist has the freedom to explore *all* the alternatives, including several YEC models, such as Lisle's anisotropic synchrony convention, Humphreys' time dilation model, and Faulkner's *dasha* theory (Faulkner 2018b). A conceptual universe allows YEC cosmologists to keep a number of theoretical irons in the fire, while at the same time providing a benchmark from which to do cosmological research at the *secular* level.

Process and development

The concepts of *process* and *development* are crucial to the argument that follows. Both ideas are derived from SFP theory as it is discussed in Chapters 4 and 5, as well as the geophysical data for the Earth discussed in Chapter 5. It is clear from Baumgardner (2000) and Faulkner (2018b) that God seems to have created the Earth in terms of an unfolding process rather than instantly. This means that during Creation Week, the Earth *matured* from an *undeveloped* state.

Applying this same logic to the universe, I propose that God assembled a mature, yet *developing* universe in His mind. This is no different than the conceptual development of Adam from Chapter 4. God designed humans to go through a natural process of maturation, starting life as a zygote, then developing into a fetus, progressing through various stages of infancy, childhood, pre-pubescence, and pubescence before becoming a mature adult. Importantly, this developmental sequence was conceived *before* a single biological cell had sprung into existence.

This logic can be applied to the universe. Since the universe is presently developing (Chapter 5), its entire developmental history, as with Adam, must have been designed and planned out before God created it. God chose to bring Adam into our existence as a fully mature human, and so it should not surprise us that God decided to do the same thing with the universe.

It is true, God could have brought an *immature* universe into our existence, causing it to develop in time and space like the "sprouting" plants and trees of Genesis 1:11 – 12 or the Earth as described in Chapter 5. Those that favor this position will most likely adopt Humphreys' time dilation model or Faulkner's *dasha* theory (Faulkner 2018b). Doing so, however, doesn't change the fact that God has every developmental facet of our universe, including its future state, fully planned out in His mind. Choosing a cosmology that matures the universe in six days merely suggests that God brought an "earlier stage" of that pre-planned sequence into our existence.

As stated above, this mature universe most likely looked very similar to the one that exists in the present, since not much movement occurs on a cosmic scale in only 6,000 to 10,000 years. Although God could have just "spoken" the entire universe into our existence in an instant, He chose instead to bring it into existence*in stages.*

According to the timeline in Genesis 1, the Earth is the first material object to be positioned in this newly created space/time reality. Thereafter, each *conceptual* object God places into this temporal context extends out *from* the Earth *into* the universe. This means that the next material objects that are "copied" and "transferred" from the realm of God's mind and "spoken" into our existence are the conceptual photons of light emanating as streams from the conceptual sun. It is not until Day 4 that the conceptual sun *itself*, and the conceptual moon and stars (celestial bodies), are brought into *our* existence along with their respective streams of conceptual light.

As an illustration, consider an entire city and neighboring countryside that loses all power enveloping everything in complete darkness. Then, in the middle of that city, within a single skyscraper, the lights begin to burn before spreading to neighboring buildings. As light flows out in all directions it illuminates the entire city, then the suburbs surrounding the city, and then the farms surrounding the suburbs. This is what it must have looked like from the phenomenal perspective given in the book of Genesis.

The Earth before the sun?

Ever since my first reading of Genesis 1, I have often been intrigued as to why God would create the Earth *first*, before even the sun. Moreover, why create the universe in six days? Why didn't God just "snap His fingers" and create the universe in an instant? The answer to those questions is deeply profound, and for those who take a young-age approach to the text, deeply satisfying.

The universe is, of course, vast. Modern scientists have been able to detect billions of stars within our own galaxy, the Milky Way, and billions of galaxies that themselves are separated by billions of light years. Immensity and grandeur on a penultimate scale are truly the order of the day. The Bible says, "the heavens declare the glory of God" (Psalm 19:1 ESV). All too often, however, this universal splendor, by virtue of its truly majestic scale, robs man of his special place in the universe. Consider Carl Sagan's view of man in contrast to the immensity of the universe:

> *Our posturing, our imagined self-importance, the delusion that we have some privileged position in the Universe, are challenged by this point of pale light [the Earth]. Our planet is a lonely speck in the great enveloping cosmic dark (Sagan 1994).*

I believe that God gave the Earth its primary position in the order of creation precisely to stop people like Carl Sagan from robbing man of his eternal value. Man is made in the image of God and is the only creature that He redeems. Contrary to popular opinion, the pinnacle of God's creation is not the universe, *it is man*. Man is God's creative, crowning glory. In order to underscore this reality, God purposefully placed the planet upon which man would dwell as first in the created order. Craig Lloyd puts it this way:

> *Notice how the earth is described in verse 2, formless and empty—* ***tohu wabohu****. The earth is without form and it is empty—there is no life in it. Then God forms and fills this formless and empty world. Days 1 – 3 deal with the changing world from formless to formed. Days 4 – 6 deal with changing the world from empty to filled. Notice that what is formed in Day 1 is filled in Day 4 and the same with Days 2 and 5, and 4 and 6. The Lord is creating a world for us. This whole creation is leading up to the formation of man on Day 6. We are the pinnacle of creation (Craig Lloyd unpublished).*

Lloyd goes on to describe another important function of the progression found in Genesis 1:

> *Scripture describes the earth in four different states. First, the world as initially created by God at the beginning of Day 1—formless and void—dark and deep. Then God begins His forming and filling on Day 1. He finishes it on Day 6 where this world is said to be very good (Gen 1:31). God sends light into the dark and separates the waters forming land. He breathes life into the world. But this is not the final perfect creation. There is day and night. There is dry land and watery depths [darkness and depths are recurring themes throughout the Bible that often describe sin and rebellion. The inclusion of "darkness" and "depths" in the "very good" but not "perfect" creation indicates the* ***potential*** *for sin]. This is not the final perfect world. There is potential for sin and turning from God. And that is exactly what happens. Man sins. The creature turns against the Creator. The world becomes a fallen world. The world is still described in terms of light and dark—land and depths. But the dark and depths do not just express* ***potential*** *for sin—they express the parts of the world in rebellion against God. Darkness and depths become symbols for death and separation from God. Genesis 3:17 calls this a cursed world. But all of this first creation looks forward to the final creation—the new heavens and new earth—where there will be no darkness and no depths. In Revelation 21 and 22—in the new heavens and new earth—there is no sea—and there is no darkness. There is no* ***potential*** *for sin or rebellion and no* ***actual*** *sin or rebellion. This is a perfect world. This is the world that all creation and all history are moving towards. We might live in a fallen world—but what we have to know—what the* ***progress of creation*** *tells*

> *us is that the final world is yet to come (Craig Lloyd unpublished; emphasis mine).*

Lloyd demonstrates a profound two-fold purpose underlying the sequential nature of the creation account. It only makes sense for God to underscore this importance by placing Earth into space/time first. It demarks this planet as intensely important to God, the reason for which is found in God's crowning creative achievement—man. A sequence that asserts the materialistic order may lead man to think himself unimportant in the universe, a reality that has been fulfilled in the words of Carl Sagan above (Sagan 1994).

Moreover, there is a redemptive-historical aspect to the creation account that anticipates the Fall and the post-Fall world, but then looks forward to the New Creation and the impossibility of further sin and rebellion due to the absence of "darkness" and "depths." Theistic evolutionary views of the creation account must acknowledge that the "darkness" and "depths" of Genesis 1 have no real *historical* basis. It is true, these words describe profound spiritual reality, but without a real *historical* background, they lose their significance in light of their real historical place in the book of Revelation.

What about SN1987A?

The possibility of a conceptual universe differs from classic mature creation arguments, in that presently unfolding cosmological events can be traced back to *real* histories. This is one of the chief complaints of the classic mature creation apologetic. Consider the phenomena of colliding galaxies. Classic mature creation arguments suggest that God created these galaxies in the act of collision. The events leading up to these collisions never *really* happened. Some YEC and most unbelievers think this is tantamount to deception, pointing out that logic and reason demand that there existed a time *before* a collision when galaxies were separated. Those who object and believe that God just created galaxies in the act of colliding must admit that those galaxies had initial starting conditions. Those conditions may not have existed in the conceptual model I am advocating, but either way, God must have at least *prearranged* their collision in His *mind*. The conceptual universe model proposed here suggests that colliding galaxies did in fact have a "history" *prior* to their being brought into the time/space historical narrative of Creation Week.

SN1987A, an exploding supernova, fits into the same category as the intertwined galaxies. The sun, Sk-69 202, that preceded SN1987A really "existed," at least in the mind of God. During its development, Sk-69 202 exploded, sending neutrinos, visible light, X-rays and gamma-rays out into conceptual space. The remnants of this explosion were "captured," as it were, during their transition from the conceptual universe to the time-historical universe of Day 4. Since God brought this mature yet developing universe into the time/historical context of Creation Week, it is only natural to assume that stars and galaxies at various stages of development were also brought into this context. This alleviates problems with Rusbult's (2003) Non-Essential Apparent Age (Chapter 2). He understands why the Earth should be created in a functional state (Essential Apparent Age) but fails to understand things like SN1987A. The model outlined here resolves these problems.

Finding the balance

YEC cosmologies exist as a spectrum of thought with two basic end members. On one side is a strict adherence to a classic mature creation apologetic. On the other side is a strict adherence to scientific methodology.

The classic approach suffers from a preconceived notion of deceit on God's part by suggesting that certain cosmological events, such as the SN1987A supernova, never actually occurred. A strict commitment to scientific methodology, although providing some good models for cosmological expansion (Faulkner 2016, 2018a, 2018b) has hitherto failed to account for the *origin* of astronomical bodies, such as planets and stars.[2] In reality, most YEC cosmologies exist somewhere in the middle of this spectrum. Advocating for a conceptual universe allows for important YEC research without limiting the potential value of secular cosmologies.

A conceptual universe may not, however, be to the liking of many YEC. This is because a process-dependent conceptual universe will most likely be void of "proof" for a 6,000-to-10,000-year-old universe. Indeed, if God brought this conceptual universe into existence *anticipating* the natural processes required for sustaining it, then we should be able to use these *same* natural processes for the purpose of understanding it.

As with the Earth, this would mean that presently active cosmological processes must advance at the same rates, *relative to each other*, as they did during Creation Week and/or in the mind of God prior to Creation Week. Non-natural anomalies should, therefore, not exist. This is not to undervalue the validity of anomalies. Secular

---

2 That is, how the constituent elements embodied within these celestial objects came together to form a planet or a sun.

scientists will often disregard and overlook anomalies because they do not fit their preconceived cosmological paradigms.

YEC should be attentive when they present themselves. Such anomalies, however, must be treated with the *utmost* of interpretive care. Given an overly ambitious commitment to one YEC cosmology over another can easily lead to speculation. This is especially true when some YEC, eager to "rescue" God from "scientific bias," jump to erroneous conclusions.

Scientific anomalies: The moon's recession

The problem

The moon is currently moving away from the Earth at about 4 cm a year. YEC have used this rate of recession as evidence in support of a young universe (Barnes 1982).

Basically, the moon affects the Earth's oceans by producing tides. Think of the moon's gravity "pulling" on the Earth as well as the oceans (Figure 2). The solid planet barely moves, but since the ocean is liquid, it bulges towards the moon.[3] Another bulge occurs on the opposite side of the planet, but this bulge can be ignored since it is not as important (Figure 2A).

Imagine that the Earth did not spin on its axis and the moon did not revolve around the Earth. In this case, the bulge would be located directly between the Earth and the moon (Figure 2A). If the Earth began to spin on its axis *at the same rate* the moon revolved around the Earth, the bulge would still be located between the Earth

3 Actually, this is not entirely correct. It turns out that most water molecules in the world's oceans are being "squeezed" or "pushed" along the Earth/moon axis, not "pulled." The model I am using here is still used in textbooks and is adequate for the purpose of clarity. This link is helpful for those seeking a more precise understanding of tidal bulges: https://www.youtube.com/watch?v=pwChk4S99i4

and the moon, almost as if the moon were "stuck" facing the bulge (Figure 2A). The situation changes, however, as the Earth's spin continues to increase *without* increasing the moon's orbiting velocity. In this case, the Earth's crust must move *through* the bulge. Think of the Earth as a paddle wheel where the paddles are represented by undulations in the crust (continents).

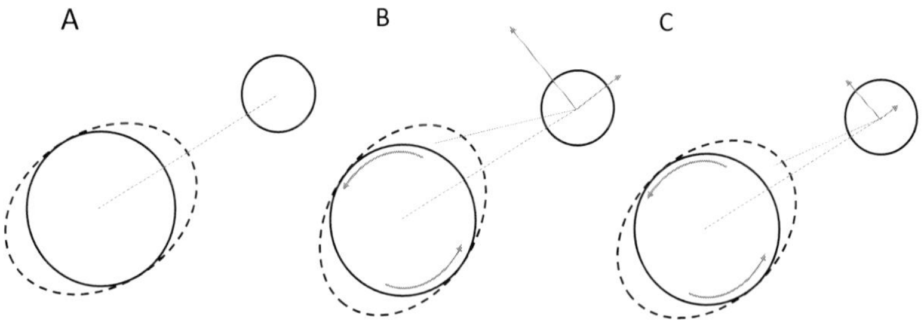

Figure 2. Rate of moon's recession changes based on where the tidal "bulge" is located. Dotted line represents two tidal bulges.

As the Earth moves through this bulge, it sort of pushes the bulge towards the east, placing it just in front of the moon (Figure 2B). When the bulge is directly between the Earth and the moon, the gravitational forces of the Earth, the moon and the oceans are all aligned (Figure 2A). Once the bulge moves to the east, however, the gravitational forces between the moon and the oceans are no longer aligned with those of the moon and the Earth. This *extra* gravitational "pull" of the moon on the bulge in the oceans produces "another" force *distinct* from that between the moon and the Earth (Figure 2B). This "extra force" "pulling" on the moon causes its orbit around the Earth to accelerate a little, while at the same time causing the

Earth's spin to decrease a little. As the moon's orbit speed increases, it also pulls it *away* from the Earth—hence recession (Figure 2B).[4]

Extrapolating this rate backwards about a billion years would put the moon close enough to the Earth as to cross the Roche Limit—the limit at which a satellite (such as the moon) can approach a planet without being ripped apart by the planet's gravitational field (Barnes 1982). Since the moon cannot have been moving away from the Earth for more than a billion years, the Earth/moon system cannot have come into existence prior to these time frames and, therefore, cannot be 4.6 billion years old. Should YEC be using the moon's recession as an example evidencing the youth of the solar system?

We will start by considering our discussion from Chapter 5. If we assume that the Earth's core, mantle, and crust were differentiated (separated) using processes such as fractionation, partial melting, and gravitational differentiation during Creation Week, then we must admit that those rates were somehow accelerated billions of times relative to the current rates at which they operate. Moreover, since secular geophysicists have been able to accurately model the interaction between these processes (Baumgardner 2000), they must have always been operating at the same rates, *relative to each other*, from the time of their inception during Creation Week down to the present (SFP theory). This means we should not find any anomalies (e.g., the absence of dense material at the Earth's center) within the differentiated Earth that would be indicative of an unknown, *supernatural* mechanism.

---

4 Technically, this is not *another* force. It has to do with torque. The angular momentum of the moon increases while that of the Earth decreases. Again, I am sacrificing accuracy for the sake of clarity.

Applying SFP theory and the conceptual universe model

For the most part, this all makes sense given the accelerated processes at work *within the Earth* during Creation Week. The Earth/moon relationship is, however, a little more complex because the moon was placed into Creation Week *after* the Earth. In keeping with a functioning, mature universe that conceptually existed in the mind of God, the Earth and the moon did interact with each other *prior* to Creation Week.

Think of the conceptual moon orbiting the conceptual Earth. Since, as with the Earth's interior parts (core, mantle, crust, etc.), we accept the interrelatedness of the solar system's parts (sun, Earth, moon, etc.), we must presuppose that the interaction between the moon and the Earth continued unabated during their transition from the conceptual universe to the time/historical context of Creation Week. In other words, the conceptual "forces," "velocities," "thermal properties," "geochemistry," etc., that were at work within the conceptual Earth/moon *system* were in *no way altered* even though the Earth and the moon were sequentially transferred into Creation Week supernaturally on different creative Days.

If, therefore, the rate of the moon's recession was the same during Creation Week as it is now, then this *exact rate* must be extrapolated *back* into the conceptual universe. Since I am proposing that all processes at work today are working at the same rates, relative to each *other*, as they did during Creation Week (SFP theory),[5] then this example would prove one of three things:

1. The Earth/moon system is about a billion "conceptual" years old. In this scenario, the rate of recession of 4 cm a year is first extrapolated back 6,000 to 10,000 years into Creation

5 Or in the mind of God.

Week. This *same rate of recession must then continue* to be extrapolated back into the *conceptual* universe (since this rate of recession remains constant relative to all over conceptual cosmological rates).

2. The rate of the moon's recession has not been constant since the *end* of Creation Week. In other words, the rate was different at some stage in the past (post-Creation Week).
3. God supernaturally originated the rate of the moon's recession along with the Earth's spin during Creation Week.

This latter option essentially rejects a conceptual universe. In this scenario, there was no initial cause-and-effect relationship between the rate of the moon's recession, the Earth's spin and other natural phenomena. This implies that unknown, supernatural mechanisms were at work during Creation Week. Simplistically, God sort of gave the Earth an initial spin and the moon an initial orbiting velocity around the Earth.[6] Choosing this option is completely understandable and biblically acceptable. As noted elsewhere, however, choosing this option mitigates origins-related model-building. This is because we have no way of testing supernatural mechanisms.

Enter our Christian geophysicist. He already believes that the Earth and the universe were created in literally six days, but he tentatively considers the possibility of a conceptual universe as well as SFP theory (thus rejecting option 3). Since he understands that all the processes that presently operate to sustain the Earth/moon system were working

6 Some may object and point out that God *knew* what orbiting velocity to give the moon and what spin to give the Earth. These rates of change were not *supernaturally* arbitrary (just give it a spin) but were chosen by God to work according to His soon to be implemented physical laws. Yet this objection is *affirming* planning on God's behalf *prior* to the creation of the Earth/moon system.

at the same rates, relative to each other, during Creation Week and/or in the mind of God, then he also believes that all these rates, including those that differentiated the Earth into core, mantle, and crust, should be somewhat consistent when extrapolating them backwards.

As it turns out, backward-extrapolation for rates of fractionation, partial melting, gravitational differentiation and radioisotope decay, all of which furnish conceptual ages of multiple billions of years (Baumgardner 2000), are *inconsistent* with the extrapolated rate of the moon's recession (furnishing a conceptual age of just a billion years or less). Our geophysicist is still not sure which of the other two options he should choose, so he carries out a literature review. It turns out that there presently exist some good, natural reasons to account for the moon's rate of recession (see Chapter 7). Given an "age" of just a billion or so years for the Earth/moon system (contrary to the majority of other "ages." See also Appendix A) in conjunction with satisfactory natural solutions that account for the moon's rate of recession, our friend should, therefore, set aside option 1 and *cautiously* select option 2 instead.

What is at stake?

Given all that has been said above, it will be important to investigate the moon's recession from a purely YEC, yet scientifically sound, perspective. Let's look at the pros and cons for accepting this piece of evidence into a YEC apologetic for a young solar system.

First, what is really gained by admitting that the Earth cannot be any older than a billion years? Yes, it "proves" that the Earth is not 4.6 billion years old, but it does not do anything to help young-Earth creationism—it in no way suggests that the Earth/moon system is only *thousands* of years old.

Second, all YEC must admit that many other rates point to the great antiquity of the Earth (fractionation, partial melting, gravitational differentiation, radioisotope ratios. See also Appendix A). Most importantly, the apologetic that jumps at an anomaly such as this without really thinking about the consequences will end up hurting the *overall* YEC position. How? Without considering all the data, the apologist is basically asserting the existence of a supernatural mechanism that was working *independently* from all the other processes at work during Creation Week.

As an illustration, let's put our geophysicist on the job doing research on the origin of the Earth/moon system. Given a supernatural-only mechanism, our geophysicist would have no option but to default to the account in Genesis, that God placed the fully formed moon into the solar system *after* the Earth. As such, he could not offer any *scientific* hypotheses as to the origin of this system, even though many geochemical relationships exist between the moon and the Earth (Baumgardner 2000). Many Christians will say that there is no need for him to offer other explanations; if the secular community does not like the "biblical" interpretation, then that is their problem. There are a few complications with this point of view.

I think most YEC would agree that our geophysicist *could* offer up *good secular* hypotheses explaining how large portions of the crust arose through fractionation, partial melting, and gravitational differentiation (Chapter 5; Baumgardner 2000). There is no need for him to default to his interpretation of the Genesis account and a billion-fold multiplication of rates, even though he believes that something like this must, in fact, have occurred. This should also be true when proposing hypotheses for the origin of the Earth/moon system. Even though our geophysicist *knows* that the moon was placed into the universe *after* the Earth, he should not be forced to default to this rudimentary explanation when proffering *secular* hypotheses to a *secular* scientific community (remember, this is his job). Ross (1994, p. 122) insightfully understands this: "Many young-universe creationists claim that star formation is one of God's miracles of creation and, therefore, could take place only during the appropriate Genesis creation day."

The *sequencing* God used to bring the Earth and the moon into our time/historical context, although historically factual, had a *theological* purpose (see above), *not* a scientific one.[7] This sequencing was a miracle, as was the accelerated rates involved in the differentiation of the Earth into core, mantle, and crust. The Earth, moon, and all their constituent parts, as well as their mineralogical and geochemical relationships are *natural*, however, *not* supernatural. These relationships, therefore, can and should be investigated scientifically because they contain scientifically meaningful information. The *sequence* that God used to place these objects into the time/historical context of Creation Week, however, cannot be scientifically investigated. It can only be *believed*.[8] Trying to find a supernatural facet that relates the Earth and the moon would be like trying to find a supernatural facet relating the constituents within Jesus' wine, bread, or fish. Given a commitment to SFP theory and a conceptual universe, one must accept that these facets do not exist. They are not meant to. The supernatural components are in the *acts* themselves, not in the objects. The acts must be witnessed first-hand for the miracles to be in any way *observed*.

---

7 Just to be sure, by "theological," I am not neglecting the historicity of a real time/space event. By "scientific," I mean that the events, although historical, cannot be tested.

8 This is also true for the supernatural components (accelerated rates) involved in the Earth's construction. This can only be seen with the eyes of *faith*.

So, for example, the disciples of Jesus knew that the wine was miraculously created because they observed the *act* itself. The rest of those at the feast had no idea that a miracle had taken place because they only saw/tasted the *object*. Supernaturally created objects will look like and retain a history that reflects their naturally made counterparts.

If YEC cannot do good, predictive science in these areas, but instead jump the gun and appeal to supernaturalism, then how can they be taken seriously in their scientific discipline? Perhaps even more importantly, how can they be taken seriously when good scientific evidence for a young Earth is forthcoming in other areas (see Chapter 7)?

Scientific anomalies: Spiral galaxies

The speed at which spiral galaxies unfold has long been used as an apologetic for a young universe. The anomaly in this case is based on the existence or nonexistence of dark matter. Without its existence, these spiral galaxies could have a life span only in the tens of millions of years—far short of the 14.6 billion years proposed for the universe. YEC have been quick to pick up on these anachronistic data, using them as evidence in support of a young universe.[9] There are a few problems, however, with this line of reasoning.

First, what is really gained by pointing out this anomaly? What's the difference if the "age" of the universe is 10 million years

---

9 Many YEC use the term "young" in YEC apologetics. To the uninformed audience, this term infers a 6,000-to-10,000-year-old universe. But when the YEC writer or speaker uses this term, what they typically mean is a universe that can be no older than tens of millions to hundreds of millions of years. This in *no way affirms*, however, that the universe is only *thousands* of years old. YEC should be careful to make this clear, especially when they present a list of examples that "prove" the "youth" of the universe. "Youth" must be defined by YEC for the sake of Christian transparency and integrity.

or 14 billion years? This only "proves" that the universe is not 14.6 billion years old.

Second, the YEC *knows* that galaxies do not come into existence in six calendar days (at least not according to rates and processes we experience today). This means they also know that these cosmological objects must have been created in a mature form retaining some kind of "apparent age" (based solely on the amount of process involved). Pitting science against science is, therefore, a poor strategy for winning people over to a YEC position. It is a bit like saying, "your view is scientifically incorrect; therefore I am right," without providing a scientific alternative supporting a *very recent* creation (thousands of years). In order to accomplish the latter, the YEC must show that *all* cosmic processes operated over periods of just thousands of years. This includes the *origin* of the cosmic entities involved in those processes (e.g., planets, moons, suns, solar systems, etc.). Since this is not possible, the YEC needs to stop pitting science against science. When it comes to Creation Week apologetics, what should be required is a robust commitment to theology and philosophy.

Third, using the spiral galaxy example for the *sole purpose* of supporting the "young" age for the universe is not a good way to *do* science. Having this kind of a mindset, our Christian geophysicist is not going to consider other hypotheses—for example the *existence* of dark matter. This means that many discoveries must be left in the hands of our friend's secular counterparts. As it turns out, more and more YEC are beginning to *support* the notion for the existence of dark matter (Faulkner 2018b). As this occurs, spiral galaxies will eventually become an obsolete example of YEC apologetics. Unfortunately, many YEC will not take these considerations seriously because to do so would mean giving up ground to the "evolutionists."

Danny Faulkner (2018b, p. 39) discusses this kind of backlash in his assessment and tentative acceptance of dark matter. One can understand why many, typically non-scientifically trained YEC will not like this newer consensus on spiral galaxies. It is one less "silver bullet."

Doing cosmological research

As stated above, there are a multitude of problems associated with all cosmological models. Notwithstanding this, I do think YEC cosmologists should be building origins-related cosmological models. Shooting holes in secular scientific theories without furnishing workable alternatives is not a good way to do science. Yet for the most part, this is what is occurring within the YEC scientific community.

Some good areas that require research relate to planet and star formation. There are plenty of models explaining the distant starlight problem (e.g., gravitational time dilation, a change in the speed of light, anisotropic synchrony convention, and mature creation or *dasha*), yet none of these models discusses the *construction* of stars, solar systems, and galaxies at the *atomic* level.

The universe is there to be understood (Genesis 1:26 – 28). God has created stars and planets with very specific thermodynamic, geochemical, and mineralogical signatures for a reason. Consider these words from Baumgardner (2000): "Of course, the tantalizing implication is that the whole solar system was formed from material of more or less uniform trace element composition." The vast physical sources of information found in the universe contain meaningful information that seems as though it can be interpreted scientifically. A commitment to a conceptual universe in conjunction with SFP theory allows cosmologists the freedom to explore this information and do good, predictive science, while at the same time remaining faithful to a literal interpretation of Genesis 1.

As a caveat, I do want to suggest the *possibility* that a scientific explanation for the formation of stars, solar systems, and galaxies *does not* exist. As I have stated elsewhere, a commitment to God's using supernatural mechanisms during Creation Week is entirely non-contradictory. God can create the universe in any manner He so chooses. As I have also said, however, such a commitment weakens scientific methodology—an accusation that YEC adopting this position must be prepared to accept.

Given the analogue of the differentiated Earth (Chapter 5), however, I *cautiously* suggest that a "naturalistic" explanation *does* exist. Some may be a little less enthusiastic, pointing out as an example the ridiculous notion of scientifically validating abiogenesis. And with this I agree. Having affirmed this, however, I do want to suggest a fundamental difference between these two acts—the maturation of the inorganic universe and abiogenesis. The development of the inorganic universe is analogous to the *maturing* of an organism. A planet starts out as an undifferentiated mass, eventually maturing into a fully differentiated "adult" planet. A child is conceived, grows into a baby, then a child, then a teenager, and finally matures into an adult. These two processes are related to *maturation*. Abiogenesis finds its inorganic analogue, *not in maturation*, but in the Big Bang—something coming from nothing.

Christians are right to *insist* that the genesis of both inorganic matter and life are miracles requiring no scientific/"naturalistic"[10] interpretation. This fact is, of course, axiomatic, since to

10 In other words, God did not use regular, rhythmic laws as He does when sustaining the universe. These two events are clearly supernatural, non-repeatable events.

date, these creative acts, creating something from nothing and life from non-life, are still best understood using an Intelligent Design model.

Chapter 7

Why Bother Doing YEC Research?

To answer this question, we first need to distinguish between the supernatural and the natural. The only way I, as a scientist, can differentiate between things that were supernaturally created (wine, almonds, bread, fish, planets, stars, solar systems, etc.) and that which has originated through normal, post-creation processes is to carefully read what God has said in His Word. If I have been *specifically* told where God has intervened in the natural order, then I must take God at His word.

This means that any scientific conclusions drawn from supernaturally created or supernaturally altered objects must *first* be interpreted through the lens of Scripture. Since the entire created cosmos was created in literally six days, I am not free to interpret the *duration* of that event any other way, even when the scientific evidence suggests otherwise.

Given this, many may wonder why Christians should bother doing YEC research at all. What's the point, some might ask, since the processes at work during Creation Week will, without the testimony of Scripture, lead to a "naturalistic"-only origin?

The heavens declare the glory of God

Scientists from all disciplines can still add to the body of YEC research using concepts in Intelligent Design. These aspects relate to both the organic-created order (the cell and information encoded in DNA) and the inorganic-created order (fine tuning of the universe). It is true that the "young Earth" aspect is missing, especially as it relates to cosmology, but young-Earth creationism is not just about "proving" the Earth is young. The universe is vast, and life is complex. God gets the glory for His creation no matter how old or how "evolved" people might think it is.

Consider a finely tuned universe necessary for anything, let alone life, to exist. Here is a direct quote from YEC Jonathan Sarfati (2015) summarizing these complexities:

- The electromagnetic coupling constant binds electrons to protons in atoms. If it was smaller, fewer electrons could be held. If it was larger, electrons would be held too tightly to bond with other atoms.
- Ratio of electron to proton mass (1:1836). Again, if this was larger or smaller, molecules could not form.
- Carbon and oxygen nuclei have finely tuned energy levels.
- Electromagnetic and gravitational forces are finely tuned, so the right kind of star can be stable.
- Our sun is the right color. If it was redder or bluer, photosynthetic response would be weaker.
- Our sun is also the right mass. If it was larger, its brightness would change too quickly and there would be too much high-energy radiation. If it was smaller, the range of planetary distances able to support life would be too narrow; the right distance would be so close to the star that tidal forces would disrupt the planet's rotational

period. UV radiation would also be inadequate for photosynthesis.

- The earth's distance from the sun is crucial for a stable water cycle. Too far away, and most water would freeze; too close and most water would boil.
- The earth's gravity, axial tilt, rotation period, magnetic field, crust thickness, oxygen/nitrogen ratio, carbon dioxide, water vapor and ozone levels are just right.

Young-Earth creationism

Young-Earth creation research with a specific young Earth emphasis is also still relevant. Since the end of Creation Week, the Earth has experienced natural rhythms proceeding at normal rates of change. This means that all large-scale[1] processes operating during this period should be interpreted using normal scientific methods of research.

Two possible exceptions exist: The Fall/Curse and Noah's Flood.

God's Curse on nature resulted from Adam and Eve's fall into sin. This Curse clearly altered the natural order in ways that would seem to impinge upon scientific endeavor (Genesis 3; Romans 8; see also Chapter 5). It may be, however, that the effects of the Curse on the *inorganic* order differed little with the processes at work during Creation Week. In Chapter 5, under the subhead, "The Fall and God's Curse on creation," I argue that although God used plate-tectonic processes to initially form the Earth, He brought these processes to a halt at its completion on Day 6. This makes sense since most natural disasters are directly or indirectly associated with moving crustal plates.

1 Excluding small-scale miracles, such as those related to Jesus' wine, bread, and/or fish.

Because of the Fall, however, God Cursed the Earth by *reestablishing* a plate-tectonic regime similar to that at work during Creation Week. This hypothesis allows for continuity between Creation Week processes and those that have been at work within the Earth since the end of Creation Week and extending to the present, excluding the geologically insignificant time represented by Adam and Eve's unfallen state. This continuity would, therefore, allow for a consistent application of the scientific method to Earth's entire geological past, avoiding the need to default to supernatural mechanisms.

The presence and/or absence of supernatural processes associated with Noah's Flood, however, is a little more speculative. It is true that God supernaturally closed the door of the ark, brought all the animals to the ark, and provided special care for Noah and the animals during their time on the ark, but did God supernaturally intervene in large-scale physical processes? Since the Scriptures are silent on the matter, it will be helpful to default to a contextual contrast.

Let's consider the differences that exist between the account of the Flood and the creation account. The latter was clearly supernatural, with large-scale geological and cosmological events occurring over *extremely* short periods of seconds, hours, and days (or occurring outside of time). Events such as those recorded in Genesis 1 simply do not occur within the providentially constrained natural order. The Flood account, on the other hand, records a large-scale geological event that occurred over a period of more than a year. This contrast suggests that God used *natural* means to generate and accomplish a global Flood. Having said that, I do think Christians should leave room for supernatural mechanisms but refrain from incorporating those mechanisms into any kind of *scientific* model.

A *firm* commitment to the workings of supernatural mechanisms during the Flood would not only be speculative but would also hinder further *scientific* research.

YEC disagree as to which rocks within the geologic column represent those deposited during the Flood event, but most agree that large chunks of that column represent pre- and post-Flood depositional environments. Since the Flood was more than likely a natural event and since all the geological events leading up to the Flood and following the Flood were natural[2] events, then the YEC should be able to apply natural, scientific methodology to the evidence when interpreting those events. This evidence, rightly interpreted, should reveal a 6,000-to-10,000-year-old biosphere encapsulated in vast sedimentary layers that themselves represent very young depositional environments.

The Coconino Sandstone and vertebrate fossil trackways

Consider the Coconino Sandstone. This sandstone formation named after Coconino County in Arizona represents a Permian sedimentary deposit that conventionally dates to about 275 million years old. In 1934, Edwin McKee, the first person to really study the formation, interpreted its depositional environment in terms of an ancient desert.

One piece of evidence that McKee used to come to this conclusion was the presence of fossil trackways in Coconino Sandstone "dunes" (Brand 1979). These trackways were likely made by a reptile or

---

2 Of course, as I have argued above, geological events associated with Creation Week, although supernatural, were nevertheless "naturalistic" in the sense that the geochemical, chemical, elemental, and sedimentological relationships that exist within these rocks should be indistinguishable from those that exist within rocks formed *after* Creation Week.

amphibian, not unlike modern reptiles and amphibians. Up until the 1970s, there had been no attempt to reinterpret these trackways or the desert understanding of the Coconino Sandstone. That is, until Dr. Leonard Brand, a YEC from Loma Linda University, published a remarkable paper documenting new trackway data that conflicted with the desert interpretation (Brand 1979).

Brand conducted several experiments with live amphibians crawling over sand-filled tanks. The experiments were conducted with dry sand, sand moistened with spray, sand that was permeated with water, and sand that was sitting *under* the water. At the beginning of each trial, amphibians could crawl over the different sands for the purpose of recording their trackways. These trackways were then compared to the fossil trackways found in the Coconino Sandstone.

The results were very interesting. The trackways created in the lab most consistent with the fossil trackways were the ones made *underwater*. Brand clearly shows that most Coconino trackways are exceptionally *distinct*, right down to the tiny toe prints (Figure 3A and B). In his experiments, Brand demonstrated that such distinct foot, heel, and toe prints were commonly made when the salamanders he used walked on the bottom of *underwater* dunes. That is right: They walked on the sand, which was *completely* underwater, leaving perfectly clear footprints. Although foot and heel prints formed in dry sand, toe prints did not, and even the foot and heel prints that did form were indistinct.

As Brand pored over many slabs of Coconino Sandstone, he also found that the trackways sometimes went "sideways." This was intensely interesting because, although the animal walked sideways *across* the dune, its toes were always pointed to the *top* of the dune (Figure 3B). Yet animals do not walk across dune faces with their toes pointing upwards. It is almost as if the animal was trying to climb the dune, but some force was pushing it across the face of the dune at the same time. Sometimes, the sideways tracks are separated by many feet. This would indicate the animal was completely picked up and transported before being dumped on the dune face some distance away. All this evidence strongly supports the notion of a very watery environment.

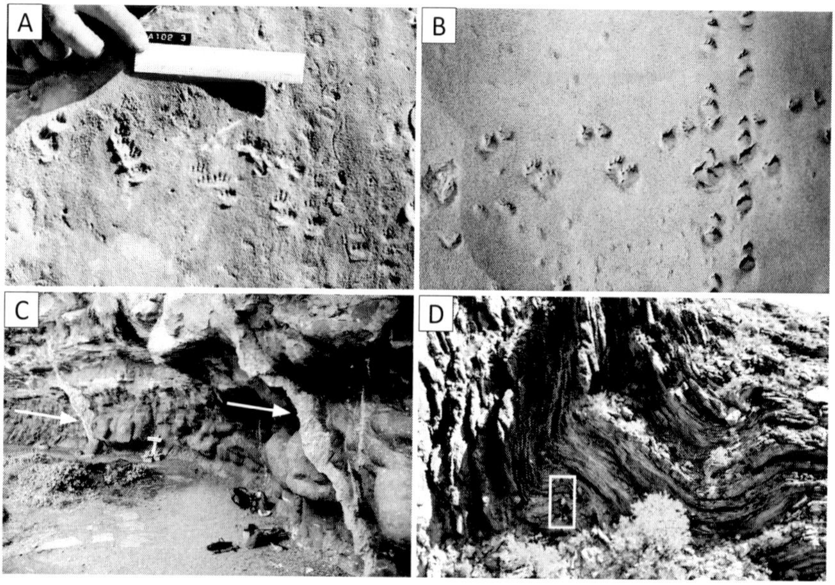

Figure 3. Vertebrate footprints found on slabs of Coconino Sandstone (A and B), sand-injectites at the base of Coconino Sandstone (C) and deformation features in Tapeats Sandstone (D). A and B. Notice the distinctiveness of foot, heel and even toe prints. In B, notice the set of tracks "walking" up the dune face. Yet how can the second set of tracks be indicative of an animal that walked **across** *the dune from left to right? Notice that the toes are pointing* **upwards**. *Animals don't walk* **across** *dunes with their toes pointed upwards. Sometimes, these "sideways" tracks are separated by many feet, almost as though something had picked the animal up and dumped it farther across the dune face. This is strong evidence for water-induced currents. Photos with thanks to Leonard Brand. C. Two very prominent sand bodies (arrows) that have been injected down into the Hermit Shale. Photo with thanks to John Whitmore. D. Extreme bending of Tapeats Sandstone. Note the person (in rectangle) for scale and the almost 90° bends. Photo with thanks to Andrew Snelling.*

The Coconino Sandstone and sand injectites

One of Brand's students, Dr. John Whitmore, also studied the Coconino Sandstone and has since published a number of papers that continue to challenge the traditional desert interpretation. One of the first papers Whitmore published interpreted what had previously been described as "mud cracks" at the base of the Coconino Sandstone in terms of sand injectites (Whitmore and Strom 2010).

Mud cracks occur over muddy terrains as moisture leaves the sediment, typically because of changes in weather. The Hermit Shale in the Grand Canyon is a 900-meter-thick sequence of shale and mudstone. According to the secular paradigm, it is believed to have been deposited by rivers and other fluvial (rivers, streams, etc.) processes more than 275 million years ago (Whitmore and Strom 2010). Over time, this flood plain is believed to have dried up as regional climate changes occurred in what at that time was a large inland area. The "mud cracks" are believed to have occurred as the area dried up.

Next came a desert, the Coconino Sandstone, which is believed to have persisted for about 10 million years and covered a large area from present day Arizona to Canada. Given the secular interpretation, it seems only natural to assume that these cracks at the top of the Hermit Shale and the bottom of the Coconino Sandstone be interpreted as *bona fide* mud cracks. Whitmore and Strom, however, were intrigued by several inconsistencies with a mud crack interpretation. Many of the cracks intruded into the Hermit Shale more than 9 meters and perhaps as deep as 15 meters (nearly 50 feet, Figure 3C). Go out to a modern-day floodplain during the dry season and you will find mud cracks that are typically less than a meter deep. These authors also found that the cracks became shallower, down to less than 1 meter in depth, the farther away they occurred from a major fault in the area—the Bright Angel Fault.

Other interesting observations included U-shaped cracks, laminations in the cracks that were vertical, not horizontal—something you should expect if sand had just fallen into the crack. Finally, they observed that the sand in the cracks was almost always pure quartz, with only a few pieces of mud mixed in with the sand—again, something that would not be true of a slow and gradual sand-filling process.

Based on these observations, Whitmore and Strom interpreted these cracks as sand injectites. Essentially, there was a massive earthquake in the area that resulted in a large amount of displacement—the Bright Angel Fault. This earthquake caused the pressurized sand, which must have still been wet, to undergo liquefaction (becoming very watery, like when you pat your foot on the wet sand at the beach). In this liquified state, and because of the weight of the overlying sediment, sand was literally "injected" down into the Hermit Shale. Of course, all these observations are inconsistent with a desert environment and are much more in keeping with a very wet one. This must also be true for the Hermit Shale since, in order for the sand to be "injected" down into the clay, it also must still have been wet and not dry as the secular interpretation suggests.

Perhaps the most intriguing aspect of this research is the relationship between the wet sand and the faulting responsible for the injection of the sand into the Hermit Shale. The sand is believed to have been deposited about 275 million years ago, yet the faulting that caused the liquefaction and subsequent injection occurred only about 50 to 70 million years ago[3]

3 This faulting is what caused the Colorado Plateau to be pushed up more than 600 meters over this 50-to-70-million-year period.

(Huntoon and Sears 1975). This, of course, presents an interesting *time* problem. How did the sand stay wet and pliable for around 225 million years? In other words, if the sand was deposited in a wet environment (itself problematic since the Coconino was supposed to represent a desert), and then other sediments were deposited on top of this environment, how did the Coconino sand remain wet enough to be injected down into the Hermit Shale after such a long period of time? The most reasonable answer is that it did not. A better overall interpretation is that the Hermit Shale and the Coconino Sandstone were deposited in very "watery" environments. (I am not going to presume these environments were related to the Flood of Noah. At this point, we will just call them "watery" environments.) Sediments from other watery environments were deposited on top of these latter two formations over a relatively short period of time (months, years, hundreds or perhaps thousands of years, but not millions). Shortly after the deposition of the last formation, the Colorado Plateau was uplifted, at which time the wet sand was injected down into the still-wet Hermit Shale.

Interestingly, the formations that lie beneath the Coconino Sandstone seem to support this interpretation. The Tapeats Sandstone, for example, is a 525-million-year-old Cambrian sandstone that was apparently deposited by an encroaching sea (Figure 4A). Since it was deposited 250 million years *before* the Coconino Sandstone (Figure 4B), and about 465 million years before the uplift of the Colorado Plateau (Figure 4D), it too should have completely consolidated into solid rock. Consequently, the faulting responsible for injecting sand into the Hermit Shale should have shattered and faulted the Tapeats Sandstone. But it didn't. Instead, the same faulting that injected sand into the

Hermit Shale caused the Tapeats Sandstone to bend like plastic,[4] sometimes at angles of 90° (Figure 3D). Of course, it is impossible for sand to stay wet and pliable for 225 million years (producing the sand injectites), let alone 465 million years (producing the 90° bend in the Tapeats Sandstone).

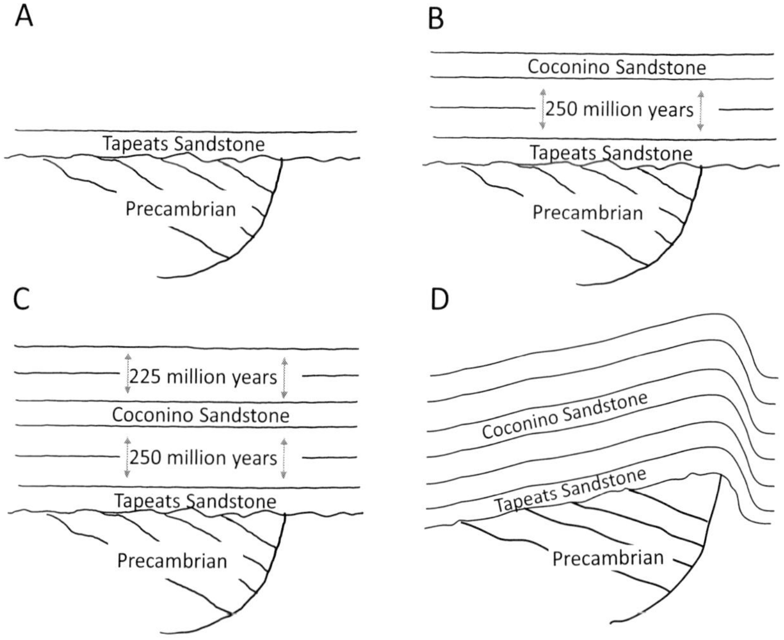

Figure 4. Depositional sequence and subsequent uplift in Grand Canyon area producing the Kaibab Monocline. A. Tapeats Sandstone is deposited 525 million years ago. B. Further deposition of several formations occurs over the next 250 million years culminating in the deposition of the Coconino Sandstone at about 275 million years ago. C. Other formations are deposited over the next 225 million years before the onset of uplift (50 to 70 million years ago). D. By about 5 million years ago most of the uplift is completed. Not to scale.

---

4 There are some faults associated with the bent strata (not shown in image). This is still not a problem since YEC believe that this uplift most likely occurred over hundreds of years. Although the sediments were initially wet, they eventually did turn into solid rock towards the end of this uplift. This produced the faulting.

The fable of presuppositionless science

Brand and Whitmore are, of course, YEC and as such have presuppositions like all scientists. They suspected that the Coconino Sandstone might have been deposited during the Flood of Noah, and they explored the evidence to see if, in fact, this suspicion was correct. Their presuppositions should not, however, be used by opponents seeking to denigrate the science behind their research.

The fact is, presuppositions cannot be divorced from scientific methodology. J. P. Moreland, a well-known and respected Christian philosopher says, "Pure, presuppositionless observations are a fable in science" (Moreland 1989, p. 64). James Clerk Maxwell's Christian presuppositions about the trinity influenced the way he viewed light: "His metaphysical picture was grounded in two assumptions: that God is a Trinity wherein each member is internally related to the others in the single whole, and that God created his world in such a way that various aspects of it, light included, reflect his triune nature" (Moreland 1989, p. 70). Moreland (1989, p. 68) even provides a bizarre example whereby scientist F. A. Kekule (1829 – 1896) developed his formula for the benzene ring after having a trance-like vision of a snake attempting to chase its own tail! With Moreland (1989), Thomas Kuhn (1996 p. 4) agrees:

> *The man who is ignorant of these fields [in this case chemistry and/or electricity] but who knows what it is to be scientific may legitimately reach any one of a number of incompatible conclusions. Among those legitimate possibilities, the conclusions he does arrive at are probably determined by his prior experience in other fields, by the accidents of his investigation, and by his own individual makeup.*

Hypotheses are not born from impartial worldviews. Everyone has a set of presuppositions that they bring to the scientific table. What matters is not the presuppositions, but a commitment to good scientific methodology. Since Brand and Whitmore acknowledge that the Flood was a *natural* event, they can and do apply the scientific method *scientifically*.

Importantly, Brand and Whitmore's work neither "proves" that the Earth is young, nor that Noah's Flood was global. The evidence is tantalizing and, in conjunction with more recent research, does seem to challenge a simplistic desert interpretation for the Coconino Sandstone (Whitmore et al. 2015; Whitmore and Garner 2018), but it would be a mistake to use these data to "prove" the universality of Noah's Flood.

Many YEC apologists make this mistake. In geology, there is a progression of interpretive thought starting at the level of individual sediment particles, then moving to the processes that deposited those sediment particles, then comparing these processes and sedimentary features to modern analogues (this is called a *facies*. A beach environment, for example, would be a specific *facies*), then comparing each facies to adjacent facies (that have their own modern analogues), then putting all these facies together to reconstruct the ancient environment (that itself, hopefully, has its own modern analogue—e.g., the zone extending from deep marine to shallow intertidal).

Once all these steps have been completed, a model can be proposed that explains the deposition of the ancient environment. YEC have yet to construct a model that fully satisfies these requirements for the Coconino Sandstone. There are many reasons for this, but for the lack of a more sophisticated explanation, it ultimately comes down to money and human resources. There are some fantastic, modern examples of underwater dune fields, but as you can imagine, extracting data from those fields would be dreadfully expensive. Unfortunately, this is one of the chief limitations in YEC research. Suffice to say, even given the virtually non-existent financial contribution that goes towards YEC research, a lot of headway has been made.

These two examples demonstrate the importance and relevance of YEC-related research. There are many scientific disciplines and fields within those disciplines that are desperately in need of YEC who can conduct good, scientific research. They can do this because all the geological and biological processes leading to the present, excluding those of Creation Week, were and are operating according to known, natural laws.

What does YEC research look like for the Christian geophysicist and/or cosmologist?

As we have seen, a commitment to a conceptual universe in conjunction with SFP theory allows the Christian who is also a scientist to conduct research and do science at the *secular* level without reference to supernatural constraints. But what about research specific to a young Earth or a young universe?

As stated above, YEC should not automatically jettison all YEC geophysical and/or cosmological models. This means that YEC are still free, for example, to use Lisle's anisotropic synchrony convention, Humphreys' time-dilation model, Faulkner's *dasha* theory (Faulkner 2018b), and/or any other cosmology the Christian thinks has merit. *Completely* committing oneself to a conceptual universe and SFP theory, however, will ultimately equate all Creation Week geophysical and/or cosmological rates, *relative to each other*, with those occurring in the present. In other words, rates of process before and after Creation Week will have the *same proportionality*. Obvious anomalies, will, therefore, not exist. Given this kind of a commitment, we will revisit Earth's decaying magnetic field and the moon's recession to see if the geophysicist and/or the cosmologist can still do *Young-Earth* Creationist research.

Earth's decaying magnetic field

As discussed in Chapter 5, current rates of electromagnetic decay clearly limit the "age" of the Earth to tens of thousands of years at most. However, instead of defaulting to the anomaly, what if our geophysicist applied solid scientific methodology to propose a *new* model? This model states that the Earth's magnetic field does indeed reverse polarity, but it does so rapidly and has been doing so since Creation Week.

Let's unpack this a little. As discussed above, the secular paradigm proposes the Dynamo Effect, whereby a decaying magnetic field is described in terms of oscillatory motion (Figure 1A). Once the field reaches its weakest point, convection currents working within the core regenerate the field strength and push the field back in the other direction.

These ideas, however, have far-reaching implications. Secular scientists already know that during reversals of the Earth's magnetic field, the Earth loses much of its protection from the sun's destructive solar winds (Wei et al. 2014). One of the consequences of such an event would include a huge loss of oxygen from Earth's atmosphere; so much so that a mass extinction event would likely result (Wei et al. 2014).

So, what's the problem? Well, there have purportedly been 183 reversals over the past 83 million years, with the actual reversal event (dipole collapse) extending over time frames that average about 7,000 years (Clement 2004). Each of these reversals should have facilitated a mass extinction event since much of the Earth's oxygen was literally sucked away from the atmosphere. Yet there is little to no correlation in the fossil record supporting such a conclusion.[5]

This is problematic from a materialistic perspective. Secular geophysicists recognize the problem associated with a rapidly collapsing dipole and want to propose rapid transitions (perhaps just a few hundred years) with long periods of stability (100,000 years or more).[6] A YEC would differ by proposing a dynamo that has rapid transitions, *but also* short periods of stability between transitions. A rapid reversal model alleviates problems associated with the current paradigm and the lack of correlating mass extinction events in the fossil record. Rapid reversals will not be as destructive and so make better sense given the fossil evidence.

Here, then, is a rich area of research from a YEC perspective, one that our geophysicist can get his scientific teeth into for the purpose of furthering young-Earth creationism.

The moon's recession

For reasons outlined in Chapter 6, it was suggested that YEC treat the recession of the moon cautiously when using it as an apologetic for a "young" solar system. Not only does this example furnish a younger "age" (a billion years) in contrast to many other geophysical and cosmological processes (multiple billions of years),

---

5 Wei et al. (2014) are wrestling with this very fact. Hence their paper that tries to correlate dipole collapse with mass extinction.

6 There are no scientific data justifying this proposal. The most logical oscillatory model proposes a uniform sinusoidal curve for both transitions (dipole collapse) *and* stability

but satisfactory scientific explanations presently exist to account for the moon's current rate of recession.

Stephenson (1981), and since then many others, have suggested that the rate of recession in the past was a lot slower than it is today based on less aggressive tidal patterns. For example, if we move the bulge in figure 2B to the west a little (Figure 2C), then this would moderate the gravitational pull between the moon and the bulge. Weakening this gravitational pull would decrease the forward orbital velocity of the moon, subsequently decreasing the rate of recession.

So, is there a satisfactory way of explaining less aggressive tidal patterns? Stephenson (1981) suggests that these less aggressive tidal patterns in the past were due to the different configurations of the continents. Using continental configurations from Permian time, for example, he proposes that the rate of recession could have been reduced by as much as one third that of the present rate (Stephenson 1981). This idea should be perfectly acceptable to YEC since most of the latter believe that the continents were in fact in different configurations before the Flood of Noah. What if our geophysicist used these data to estimate the rate of past plate movement instead of defaulting to an anomaly that can be explained only supernaturally? Is it possible that *rapid* emplacement of continents could be deduced from these data? Wouldn't this be a good area of research?

Putting on the "theological glue"

Adhering to this book's synthesis does not remove scientists who are also Christians from the world of YEC research. On the contrary, it promotes a solid scientific methodology that must be adopted if young-Earth creationism is to have any real impact, especially in the scholarly and academic realm. Of course, as with any kind of "evidence," there is always an aspect of ambiguity that will lead one down different interpretive paths. Everything said in this chapter can be interpreted in a number of ways, and more often than not, that interpretation usually depends on the presuppositions of the individual. The next chapter discusses this ambiguity in terms of a Christian philosophical and theological perspective, providing an important facet to this book that acts like a glue binding it all together.

Chapter 8

Theological and Philosophical Implications of SFP Theory

Shouldn't the creation abundantly testify to the God of Israel as Creator?

One of the most misunderstood objections to sound mature creation apologetics is the conviction that God would not lead us astray on *theological* grounds. This objection is derived from passages such as Psalm 19 and Psalm 104, which point humans heavenward for the purpose of glorifying God in creation (Ross 1994).

Another similar passage that is often likewise referenced is Romans 1:19 – 20 (ESV): "For what can be known about God is plain to them, because God has shown it to them. For his invisible attributes, namely, his eternal power and divine nature, have been clearly perceived, ever since the creation of the world, in the things that have been made. So they are without excuse." Jon Greene (2012), an OEC, says this: "It is not in God's nature to create something young and fallaciously give it the appearance of age. The concept runs contrary to Romans 1:19 – 20."

Objections such as this, however, misunderstand the very purpose of God in redemption. Psalm 19, 104 and other such passages are mainly concerned with a broad theological fact—God is powerful, and this power is displayed in the created cosmos.[1] Creation cannot, however, lead mankind to worship the *true* God. This is Paul's point in Romans 1:19 – 32. Certainly, man can *perceive* that a great and powerful God exists, but the creation will not lead him to a true and saving knowledge of that God.

Douglas Moo (1996 p. 105 – 106), a prominent and highly respected evangelical scholar, says this in his commentary on Romans: "For Paul here [verse 20] makes it clear that 'natural revelation,' in and of itself, leads to a negative result." And, "But it is vital if we are to understand Paul's gospel and his urgency in preaching it to realize that natural revelation leads not to salvation but to the demonstration that God's condemnation is just: people are 'without excuse.'"

In conjunction with the conscience, the knowledge of God through the rest of the created order leads the unbeliever to the conclusion that absolute right and wrong exist. Consider Paul's words: "They show that the work of the law is written on their hearts, while their conscience also bears witness, and their conflicting thoughts accuse or even excuse them" (Romans 2:15 ESV). This verse is incredibly instructive; Paul is saying that those who have never even heard about the God of Israel or Israel's Law still have a conscience that,

---

1 Well known OEC John Collins agrees. See his chapter, "How to Think about God's Action in the World" in *Theistic Evolution: a Scientific, Philosophical and Theological Critique* (Moreland et al. 2017, p. 671).

although flawed, nevertheless furnishes each person with a standard of right and wrong.

Even this flawed, and obviously deficient, standard, however, cannot be perfectly obeyed. This means all these pre-Christian-era unreached people are accused and subsequently condemned based on *their own* standard of right and wrong. This is why Paul can say in Romans 3:11 (ESV): "no one understands; no one seeks for God." The only thing that will bring a person into a saving knowledge of Jesus Christ is *special revelation* and *faith*. Moo (1996, p. 107) agrees: "At the very center of every person, where the knowledge of God, if it is to have any positive effects, must be embraced, there has settled a darkness—a darkness that only the light of the gospel can penetrate."

General revelation, then, brings only a sense of condemnation (Moo 1996) readying the unbeliever for the gospel that, when believed, halts the increasingly insensitive conscience from spiraling into greater depths of depravity (Romans 1:20 – 32). Once understood, this theological perspective eradicates any notion that general revelation is in some sense *on par* with special revelation (Ross 1994, p. 57) or that it can *accurately* describe thc *origin* of the created order (*contra* both Ross 1994 and Greene 2012).

A great, powerful, eternal Creator can surely be understood from nature, but unless the gospel of Jesus Christ regenerates the heart of those witnessing that creation, they will consistently choose to rebel against the true God of creation, choosing rather to worship the creation (idolatry and false religion). Once the gospel is believed and the heart regenerated, the Christian will witness the same creation as they did prior to their conversion, but now, through the power of the Holy Spirit in conjunction with God's revealed Word, the believer can rightly interpret its origin (Hebrews 11:3).

Ambiguity and room for unbelief

Freedom to disobey God has been decisively built into the ebb and flow of His special revelation for the purpose of exposing unbelief. Consider the account of Adam and Eve's fall into sin and rebellion. God created our first parents and provided for their every need but sternly warned them not to eat of the fruit of the tree of the knowledge of good and evil. Why put this bad tree in this very good creation? Obviously, without the freedom to disobey, Adam and Eve would never have the freedom to truly obey. God created humans with free will and so the inclusion of the tree served to cultivate this human attribute.[2]

Often, this *freedom to disobey* comes in the form of purposeful ambiguity. Consider the concept of a suffering servant, a notion entirely coherent to the modern-day Christian but not understood by the Jews of Jesus' day. Even the disciples did not understand this until *after* Jesus had risen from the dead (Matthew 16:23). Why did everybody get it wrong?

Although the incipient facts about a suffering Messiah were apparent within the Old Testament (Genesis 3:15; Daniel 9:26; Isaiah 53; Zechariah 12:10), its primary prophetic focus embodied the theme of a conquering King. Yet in his first advent, Jesus was nothing more than a common carpenter. His body did not emanate light. Neither did he go about on brilliant silvery clouds from which he hurled down thunderbolts of lightning. Jesus' first coming was very *human*. This humble display of humanity was a constant catalyst

---

2 I like to read the Scriptures and interpret them as the authors intended. This often means not imposing my theology on the text. I am unashamedly reformed in my soteriology, but I will not shy away from those texts that seem to appeal to human freedom. The argument that follows may "sound" non-reformed but is not. It is merely free from a theological imposition that is foreign to the text and the ideas found in the text.

for ambiguity as to his *real* identity and, in part, helps explain many of his teachings, including his use of parables.

Many Christians mistakenly think that Jesus' parables were meant to simplify the message of God's kingdom for the purpose of communicating a spiritual reality. In some sense this is true, but only for the person who was *drawn into the story*. Jesus clearly told us what function parables served: "To you it has been given to know the secrets of the kingdom of God, but for others they are in parables, *so that* 'seeing *they may not see*, and *hearing they may not understand*'" (Luke 8:10 ESV; emphasis mine).

The parables were a form of judgment based on the consistent unbelief and petulance of the crowds, which stemmed from a disdain of Jesus' very *human* presence. In conjunction with Matthew 7:7 (ESV): "Ask, and it will be given to you; seek, and you will find; knock, and it will be opened to you," it becomes apparent that Jesus requires *faith* to move past what at first seems confusing (his very human presence instead of a glorious conquering King). If the crowds chose not to seek, then instead of the Messiah, they would merely see a disturbing man who spoke of eating flesh and drinking blood (John 6:56). If they chose not to knock, they would see a man controlled by Satan rather than the Son of God who had power over Satan (Mathew 12).

Jesus Christ's glory was only *partially* revealed in his first advent. A full revelation must await a yet future period (1 Peter 1:13). Miscalculating Jesus' identity based on the ambiguity of Jesus' first advent was, and still is, a big mistake. This is the antithesis of faith. Jesus blessed those who believed, *despite* a clear revelation of his divine identity, which was most often displayed in his miracles (John 20:28). Jesus' life *in the flesh* was one of humble ambiguity that required the eyes of *faith* to see him for who he really was—the all-powerful Creator. Ambiguity then, seems to be a tool used by God to tease out faith *as well as unbelief*. This makes sense given Hebrews 11:6 (ESV): "And without faith it is impossible to please him, for whoever would draw near to God must believe that he exists and that he rewards those who seek him."

Faith is required *first*. The reward comes *subsequent* to faith.

Ambiguity and God's creation

I believe that this same kind of ambiguity has been *purposefully* built into the creation. This is not a novel idea. Wise (2002, p. 12) says this:

> *So God has placed enough flexibility in His creation to permit us not only to reject truth but to create, accept, hold to, and act out non-truths ... We might even say that ambiguity could actually be built intentionally into the very nature of creation.*

This concept contradicts a common, evangelical belief that presupposes God's creative strategy should be *obvious* in the things that He has made. This is true of many in OEC, TE and even YEC communities.

Consider the words of Rusbult (2003): "IF God wants to avoid misleading us with false history, then He will create a universe that is old, so it can actually be the age it appears to be" (emphasis his). Not only is there no theological basis for this mindset (see above), it contradicts the way in which God has chosen to bypass conventional human reasoning.

Consider Paul's words in 1 Corinthians 1:21 (ESV): "For since, in the wisdom of God, the world did not know God through wisdom, it pleased God through the folly of what we preach to save those who believe." This passage is rich with redemptive truths. Note first that in "God's wisdom," which is obviously perfect, "the world did not know God through wisdom." The latter wisdom refers to man's fallen, flawed, unredeemed and foolish reasoning *without* reference to God. In other words, it is what man can know about reality in and of himself, without reference to *special* revelation.

In short, God is saying that He has decreed (in His wisdom) that man cannot come to know Him (have a true and saving knowledge) by appealing to his own intellect and reasoning. Now, the antagonist will say that all humans use reason to interpret and understand the very words of Scripture, so reason in and of itself is not evil. And with this I agree. The problem comes when two conflicting *reasonable* conclusions exist.

Applying this to the natural order, there is much in science, as we have seen, that contradicts God's revealed Word in Scripture (Chapter 2). Reason that is worldly denies what God has said for the sake of what *man thinks* is reasonable. Yet a mature created order is, of course, *unreasonable* to man's fallen nature. 1 Corinthians 1:21 is essentially saying that by God's decree, man's reason *as it interfaces with creation* cannot lead to ultimate Truth. We should not expect, therefore, the created universe to contain theologically salvific arrows that point people to the true and rightful Creator of the universe. The *only* way for man to come into a relationship with the living and true God is by *believing* in the *foolishness of the cross*—reason that is applied *solely* to God's revealed Word. General revelation can lead to the knowledge of a powerful Creator, but it cannot lead one to know *who* that Creator is. *Only* special revelation can do that. This is not deceitful; it is the way God has chosen to eliminate man's pride.

Considering that all this is God's design, why is it that young-Earth creationism is viewed as a stumbling block to evangelism by theistic evolutionists? It is because they know that a YEC perspective seems irrational to their materialistic peers. What TE are inadvertently doing, however, is putting away the *foolish* things of God for the sake of exalting the *wise* things of the world. Yet they are missing God's design. The ambiguity in nature is *meant* to promote both belief *and* unbelief. It is the humble fool who chooses God based on reason plus special revelation.

Given this, I think it makes perfect sense for God to construct a universe *without physical evidence* of the miracle that brought that universe into existence. As I suggested earlier, God could just as well have created a universe where every newborn baby came with the words "Made in Heaven by the God of Israel" stamped on their buttocks. Could He not have done this? This would have made things *unambiguous*, right? But this is not how He did it. God has ordained that mankind accurately perceive reality, including its origins, *through the lens of Scripture*. The creation account must be *believed*, not evaluated by modern scientific methodology (human wisdom).

This does not mean, of course, that the universe is devoid of evidence for God. It is full of evidence and truly declares His power and His divine nature. This can be seen at the macro-scale in the intricate, mathematically wondrous, and symmetrically consistent nature of the universe, as well as at the micro-scale in the complex, informationally perfect, and functionally intelligible nature of DNA. For those who look with eyes of faith, there are also many "evidences" that suggest life on Earth is indeed very young.

Some may disagree, but the fact is all evidence can be interpreted in several ways. There is room for unbelief. Without divine, special revelation, individuals are free to interpret the creation in any way they so desire. But as we have seen above, this "freedom" will *consistently* lead to a rejection of the God who designed it: "they exchanged the truth about God for a lie and worshiped and served the creature rather than the Creator" (Romans 1:25 ESV). God requires that such individuals stop trusting in their own wisdom, exercise *faith*, and start trusting what God has said *in His Word*, even if that means believing things that *seem* foolish. The ambiguity is itself a check on our pride.

Ambiguity as a means of grace

Purposeful ambiguity, in both general as well as special revelation, is actually a means of grace to unbelievers. This may sound contradictory but consider Jesus' words in Matthew 13:12 (ESV): "For to the one who has, more will be given, and he will have an abundance, but from the one who has not, even what he has will be taken away." Luke (8:18 ESV) adds these sobering words in the parallel account: "Take care then how you hear."

A holy God, by virtue of that holiness, must react to rebels who knowingly and intentionally sin in God's presence. The account of Nadab and Abihu from Leviticus 10 serves to illustrate this principle. In chapter 9 of Leviticus, as part of the official inauguration of the Levitical sacrificial system, God promises to show Himself to the people of Israel (Leviticus 9:4). In verse 24 we read: "And fire came out from before the Lord and consumed the burnt offering and the pieces of fat on the altar, and when all the people saw it, they shouted and fell on their faces" (Leviticus 9:24 ESV). In the very next verse (chapter 10:1), right after God had powerfully and supernaturally revealed Himself, we read the sad account of Nadab and Abihu's sin. It appears Nadab and Abihu had gotten drunk on the job (verses 8 and 9), leading them to make a "strange" offering (verse 1). The consequences were severe and immediate: "And fire came out from before the Lord and consumed them, and they died before the Lord" (Leviticus 10:2 ESV). Then God says this to Moses: "Among those who are *near* me, I will be sanctified" (Leviticus 10:3b ESV; emphasis mine).

This was a special inauguration of a solemn event (the institution of the Levitical priesthood) where God powerfully and supernaturally turned up. If that was not enough, Nadab and Abihu were priests of God called to an extremely high standard of conduct. The entire account sheds light upon the nature of a holy God as He deals with sinful humans. The closer a sinful human gets to God's manifest holiness, the more accountable that person becomes for his or her actions.

We are all responsible for the revelation we receive. Rejecting or rebelling against God's revelation, whenever it is presented, results in increasing spiritual darkness. This is a form of judgment. Miracles and other supernatural events represent *full revelation* from God. Rejecting or rebelling against the latter is not taken lightly. This was a hard lesson for Nadab and Abihu.

This same theme permeates the teachings of Christ in the New Testament. Rejecting Jesus' words and actions is one thing (Mark 3:28) but rejecting his identity after being exposed to multiple miracles is another (Mark 3:29). The miracles of Jesus verified his identity, not only as the Messiah, but as God incarnate. If the sole purpose of life is to acknowledge and worship God, and that purpose is rejected after full revelation (miracles), then there is no further purpose to life. All that is left is judgment.

That humans might consider rebelling against their Creator given unambiguous "proof" as to His identity might seem incredible, but it happened. The people of Israel, for example, consistently rebelled against God even after He powerfully and supernaturally delivered them from Egypt (Deuteronomy 9:7). The result was devastating. Ananias and Sapphira lied to the Holy Spirit, even after witnessing amazing miracles at the hands of the Apostles (Acts 2:43). This led to instantaneous death (Acts 5:1 – 11). Many of the Jews did not believe in Jesus even after Lazarus was raised from the dead (John 12:9 – 10). Jesus said: "If they do not hear Moses and the Prophets, neither will they be convinced if someone should rise from the dead" (Luke 16:31 ESV).

Most people think that automatic obedience to God will result given *enough* evidence. Some people say, "If God just came and touched me on the shoulder, then I would believe!" But according to the testimony of Scripture, this is not true. The problem with unbelieving humans is not whether or not supernaturalism is *real*, as though a sign indicative of the supernatural will somehow change minds. The problem is at the very heart of man, who, according to Paul, hates (the *true*) God (Romans 3:11). One cannot come to know God through a supernatural experience. God must be *trusted*, and in order to trust God, one must come to know what He is like and how He interacts historically with humans—hence a Bible with historical narratives.

In Hebrews 11:6 (ESV), the writer puts it this way: "And without faith it is impossible to please him, for whoever would draw near to God must believe that he exists *and that he rewards those who seek him*" (emphasis mine). It is not just about believing in God's existence. Humans must also believe that He is *good* (a rewarder of those who seek Him). Many will object to this latter requirement, insisting that God is not good. How, for example, can a good God condemn people to eternal punishment for minor offenses? Or how can a good God allow bad things to happen to good people? And, of course, this is precisely why such people will not believe, *even* given a supernatural visitation. It has nothing to do with God's existence and everything to do with His *perceived* character.

Approaching the Scriptures (historical accounts of God's interactions with humans) with humility instead of hostility is essential as a first step towards genuine faith. Veiled ambiguity as to God's existence is, therefore, a purposeful measure of God's grace towards His enemies. An unveiling of His full glory towards hostile subjects will only result in rebellion followed by swift justice—a consequence of rejecting full revelation.

Given all that has been said above, imagine the created order everywhere stamped with the miraculous, where daily cycles of *supernatural* events (for example, angels flying back and forth) instead of natural ones governed and sustained the physical universe. A world that functioned in this way would mean the end of faith, where willful unbelief resulted in severe judgment. Reynolds (1994, p. 463) seems to be tracking along similar lines: "If we knew in some strong sense that [young-Earth] creationism was true, then only the utterly irrational could choose to reject God." Of course, even Reynolds seems to miss the outcome of such a reality. The fact is that people would still choose to rebel against God, even given a "strong sense" of a purely supernatural creation:

> *The rest of mankind, who were not killed by these [supernatural] plagues, did not repent of the works of their hands nor give up worshiping demons and idols of gold and silver and bronze and stone and wood, which cannot see or hear or walk, nor did they repent of their murders or their sorceries or their sexual immorality or their thefts (Revelation 9:20 – 21 ESV).*

The natural revelation, as it presently exists, is not sustained by a miraculous set of laws. It is natural by design. Even so, this natural universe is *enough* to convince man about the existence of a powerful, good God. This natural revelation, however, does not result in salvation. Rather, it results in condemnation and preparation for the gospel (Romans 1:19 – 20; Moo 1996). Rejection of God based on the witness of God in the natural revelation will not lead to instantaneous, severe judgment since supernaturalism (full revelation) is absent. This rejection will, however, lead to more and more spiritual darkness (*mitigated* judgment), ultimately ending in final judgment at death when the grace offered in life expires.

Masking *full revelation* from humanity is thus a gracious act of God that enables unbelievers to reject God as revealed in Scripture and yet enjoy a mature, functional universe in which to live and find meaning. Since, however, the unbeliever has rejected God's *special* revelation, they are led to interpret the *maturity* of the created cosmos in terms of natural processes operating over billions of years. This is the natural inclination of their heart. Yet even here there is grace. The apparent history of a mature universe allows the unbeliever to pursue that history, literally, to the ends of the universe and the beginning of time. The believer, however, knows that this history is *apparent* because they have believed God's Word. For the unbeliever, this apparent history becomes a form of grace because they have another reality in which to place their trust if they reject the one God has proposed.

The importance of faith

Many Bible passages point to the requirement of faith. In Hebrews 11:3 (ESV) we read: "By faith we understand that the universe was created by the word of God, so that what is seen was not made out of things that are visible." But how is it that we *understand* these things about the universe? Obviously, the writer of Hebrews is drawing our attention to the *creation account*, calling us to *believe* what God has said in Genesis. Theistic evolutionists have sadly missed this. They insist that God allowed the ancient Hebrews to borrow pseudo-scientific ideas from their pagan neighbors and incorporate those ideas into the text of Genesis (Lamoureux 2010). Yet if this is true, in what sense can we believe, trust, and have *faith* in the creation account *as it has been revealed* (remember, the writer of Hebrews is calling us back to Genesis)? In what sense can we believe that God brought forth the universe from unseen things in a sequential series of six creative days?[3] We obviously cannot.

Let us have a look at some other aspects of faith. In Hebrews 11:6 (ESV) the writer says, "Without faith it is impossible to please God." This statement just makes sense. How many times have we questioned God's goodness, integrity, power, knowledge, wisdom, fill-in-the-blank, because we simply did not have all the facts? Many times, not always, we come to see God's wisdom work itself out in our lives. Yet unless we have a vantage point from which to know all facts, all knowledge, all wisdom, across all time and all space, we will *always* be subject to faith (trust).

Those who deny this are simply being proud—"I won't believe it until I see it." Now, I am not suggesting that naturalists are exempt from faith when making assumptions about events that occurred in the distant past. Everyone has some measure of faith because they

---

3 Appealing to the Big Bang cannot be incorporated into this text. The writer to the Hebrews is not thinking of the Big Bang, a singularity or anything related to modern cosmology. He is asking us to have faith in the account *as it is revealed in Genesis*. He is calling us *to the Bible*.

were not there to witness the event. Even so, I do believe it is the Christian who must exercise *more* faith. That is because the materialist rests on his daily experiences to interpret past events and predict future ones. Every day that he is alive, the unbeliever's convictions about natural laws harden his heart towards the reality of a supernatural Creator. Once this initial set of experiential presuppositions are set in place, it does not take a lot of faith to extrapolate these presuppositions back into time, especially when one considers that the universe God has made is a decidedly natural one.

Finally, I am not suggesting that "evidence" is unimportant. The resurrection event was designed to be *witnessed*. The miracles and the Old Testament prophecies were all crafted for the purpose of bearing *witness* to the God of creation and the plan of redemption. God has not left us without evidence, nor does He want prospective Christians to "take a blind leap of faith." The Christian religion is unquestionably an evidence-based faith in the sense that God has furnished *enough* evidence to satisfy the intellect. There must, however, be a balance between *evidence* and *faith*. Finding that balance is not always easy, but it is usually obvious when Christians swing too far to the left or to the right of it.

Chapter 9

Is God Lying?

Do supernaturally created objects retain *deceptive* natural histories?

The most serious objection to a sound mature creation apologetic is related to deception on God's part. Consider Rich Deem's (2013) words:

> *Here is where the appearance of age argument distorts the reality of God's creation. Since we know that objects in the universe are 13 billion years away, it must have taken the light 13 billion years to reach us. Therefore, the universe must be at least 13 billion years old. Either God created the universe at least 13 billion years ago or He deceived us by making it seem to be 13 billion years old.*

This is a good place to revisit some of the biblical examples discussed above to test Deem's statement: "Either God created the universe at least 13 billion years ago or He deceived us by making it seem to be 13 billion years old."

Consider Jesus' wine. If we could somehow procure a sample of Jesus' wine and submit it to a series of experiments in a modern scientific laboratory, what kinds of results might we expect as to the wine's origin? Without revealing its true provenance to a secular scientist, he may conclude that the grapes had been grown in a particular climate, at a particular geographical location, and under the stewardship of a particular winemaker over a certain period of extended time. At the very least, he would assume that a vine was grown, grapes were harvested and crushed, and juice was fermented. In other words, he would conclude that the wine retained a *natural* history.

Now assume that the scientist was told that this wine was created instantly from water. Is God deceiving the scientist? No. That is because the origin of the wine was *made known to him*. Neither could God be accused of deception if the same set of conditions were applied to the almonds, the snake, the bread, the fish, new organs, cells and skin. When God says that He has acted in nature to create objects supernaturally, humans are not given the option to interpret their origin in terms of materialistic processes. Moreover, if humans decide to forego God's revelatory interpretation, they will undoubtedly discover an *apparent* origin that *will* lead to erroneous conclusions. Yet this is not deception on God's part because He has made the necessary *historical* origin abundantly clear.

Consider the miracle of the incarnation and the biological constituents of Jesus' DNA. Jesus is fully God, but he is also *fully human*:

> *For this reason he had to be made like them,* ***fully human in every way****, in order that he might become a merciful and faithful high priest in service to God, and that he might make atonement for the sins of the people (Hebrews 2:17 NIV; emphasis mine).*

We should assume, therefore, that Jesus would have diploid somatic cells (two sets of chromosomes, one from the father and one from the mother), just like all other humans. If not, then we cannot say Jesus was *fully* human. Consequently, if our secular scientist could run a test on Jesus' blood, the results would almost certainly indicate a chromosomal blueprint that was biologically indistinguishable from any other human (i.e., having two sets of chromosomes). For those who look at things only in the natural, this would indicate that Jesus was nothing more than a man with a *biological past* just like everybody else. If you told the scientist that this DNA was from a man born of a virgin, he would obviously think you were mad.

Does this mean God is deceiving us? Of course not. Why? Because He *told* us about Jesus' virgin birth and divine nature, *despite* what the biological evidence suggests. Those who believe God is lying because the universe was created using supernatural formative processes must also affirm that He is lying about the SFP (an extra set of chromosomes) woven into Jesus' biology. The TE and/or OEC may object to this by stating that Jesus' blood is not available for testing, which means we do not know *for sure* that he had diploid DNA. At the very least, however, they must admit that Jesus' Y chromosome was created using SFP. Jesus is a male and his mother was a female—only passing on an X chromosome. That Jesus' DNA is not accessible is simply a bad argument.

Biological science tells us that Jesus, at the very least, has a Y chromosome, but most likely was fully diploid. This is a *scientific fact*. So here is my question for those who think God is deceiving us about the use of SFP: On what basis do they believe Jesus was *virgin born*? If the answer is in any way related to God's *special* revelation, then they are taking God at His word *despite* what the scientific evidence suggests. Why are they willing to believe in the supernatural history associated with Jesus' DNA but not the supernatural history woven into the universe? This is especially important since rejecting the reality of Jesus' supernaturally created partial genome will affect one's eternal destiny.

The deception of modern Gnosticism

Claims of deception on God's part work both ways. If deep time and Darwinian evolution represent God's creative strategy, then why did He use very transparent language to describe a creation that was brought into existence over literally six days? TE counter this claim by deferring to a faulty view of accommodation–i.e., the purposeful compromise on God's part that allows for the erroneous inclusion of the original writer's faulty science into Scripture.[1] Since these "simple" people did not have a grasp of "modern" evolutionary science, we should not expect these views to be incorporated into the creation account. Consider these words from TE Denis Lamoureux (2010):

> *It is important to note that ancient peoples also understood the origin of life from an ancient phenomenological point of view. Biological evolution was not even a consideration because in the eyes of the ancients, hens laid eggs that always produced chicks, ewes only gave birth to lambs, and women were invariably the mothers of human infants. Living organisms were therefore static and never changed. In conceptualizing origins, they used these*

1 God's accommodating finite human minds is, of course, a theological reality, but there is a big difference between accommodation as traditionally understood and the "modern" version espoused by TE. For example, if your child comes to you and asks where baby came from, you are not going to tell her that a stork dropped baby off at your door in a little folded handkerchief. Instead, you might say that baby grew inside mummy's tummy and when it was time for baby to enter the world, mummy went to the hospital where the doctors helped baby arrive. This simple account is accommodation *without* neglecting scientific accuracy and/or historicity. TE want us to believe that Genesis 1 consists of stork stories.

> *day-to-day experiences and retrojected (to cast back) them to the beginning of creation. Ancient peoples came to the very reasonable conclusion that life (and the universe) must have been created quickly and completely, "after their kinds" as stated ten times in Genesis 1. Termed "***de novo*** *creation" (Latin* ***de:*** *from;* ***novus:*** *new), this was the best origins science-of-the-day. It appears in most ancient creation accounts and it involves a divine being/s who acts rapidly through a series of dramatic interventions, resulting in cosmological structures and living creatures that are mature and fully formed. With this being the case, it becomes evident that the God-of-the-gaps model of divine creative action is ultimately rooted in* ***de novo*** *creation, an ancient origins science.*

"Accommodation" is summarized by A.N.S. Lane as follows: "God speaks to us in a form that is suited to the capacity of the hearer" (Poythress 2019, p. 324). Importantly, up until the enlightenment, this traditional view of accommodation took care not to deny the truthfulness of Scripture (Poythress 2019, p. 325). TE have clearly crossed this line (Lamoureux 2010). Accepting an erroneous view of Scripture, however, is only part of the problem. Poythress (2019) perceptively illuminates some of the perils of this more "modern" consensus by using the analogy of a father who uses simplistic language when instructing his son for the purpose of conveying truth with clarity. TE see themselves *not* as the son, but as an on-looker, on the *same intellectual level* as the father. In other words, they have a *privileged* position and can thus appreciate what the father (God) is "leaving out" of the dialogue with his son (humans). The problem of course is obvious. By placing themselves in this privileged position, they no longer think of themselves as the intelligently challenged son. Essentially, they have become modern Gnostics (Poythress

2019, p. 327). Poythress says it well: "[those who hold this view try] to transcend our human limitations in order to watch God over his shoulder, so to speak, and thereby to know the ways in which [they] ... can and cannot receive Scripture at full value" (2019 p. 328). In short, those espousing these views have a claim to some kind of "special knowledge."

Carrying the father/son analogy further, Poythress discusses another common pitfall with the modern consensus that he describes in terms of "underestimating divine power" (Poythress 2019, p. 330). We, as sinful and fallen humans, can sometimes get frustrated when trying to communicate a difficult concept to our children. The temptation is to place God in a similar box; God, like a human father, gets frustrated because he cannot find a way to communicate to his "simple-minded" son. But as Poythress points out, God is not limited by the "circumstances." Surely, man cannot "frustrate" God's desire to communicate truth! (Poythress 2019, p. 330). As Poythress says: "Precisely because God is the absolute creator, human finiteness offers no resistance, no problem, for communication" (Poythress 2019, p. 330).

This becomes particularly relevant when trying to decipher the origin of the creation, the universe, the biosphere, and man. If God wanted to convey man's evolutionary history over millions of years, He surely had the ability to do so. This is especially pertinent given the capacity of ancient cultures to communicate simplistic versions of modern evolutionary theory (Contra Lamoureux above).

Poythress (2019 p. 193 – 194) describes, for example, the *Atrahasis Epic*, that has humanity originating from seven pairs of humans (not one) that involved an extended process operating through multiple stages. The Korean legend of Dangun contains a part in which a bear becomes a woman. A Tibetan myth says that the "Tibetan people originated from the union of a spirit/ogress with a monkey or ape."

The Greeks and the Romans likewise had similar ideas. The following excerpts are from a University of California, Berkeley, Museum of Paleontology article (1996):

> *Evolutionary theory begins with the Ionian philosopher Anaximander (ca. 611 – 546 B.C.E.) ... Humans were not present at the earliest stages; they arose from fish ... Another Greek philosopher, the fifth-century materialist Empedocles of Acragas (in Sicily), Empedocles claimed that the Earth had given birth to living creatures, but that the first creatures had been disembodied organs. These organs finally joined into whole organisms, through the force of Love ... Much later, the Roman poet and philosopher Titus Lucretius Carus (99 – 55 B.C.E.) wrote his long philosophical poem* **De Rerum Natura** *("On the Nature of Things"). In this poem Lucretius proposed, among other things, an "evolutionary" theory similar to that of Empedocles ... Here again, species were born out of the Earth, formed by the chance combination of elements. Natural selection led to the extinction of once-living "monstrous" organisms. Those organisms that survived either survived because of their strength, speed, or cunning, or because of their usefulness to people.*

Multiple cultures, including the Greeks and Romans, clearly held to materialistic elements found within today's modern scientific worldview. Although the Hebrews were geographically separated from some of these cultures, this does not necessitate their being oblivious to such views, especially since ancient peoples were interested in the worldviews of other nations. Given this, and God's concern for other nations in the Old Testament (Isaiah 42:6), it seems implausible that God would have allowed the ancient Hebrews to use an inaccurate scientific perspective (*de novo* creation) when more "precise" versions were easily accessible.

In summary, TE counter a literal, calendar-day reading of Genesis 1 by appealing to a faulty view of accommodation. This appeal fails, however, because incipient evolutionary ideas were common to many ancient cultures and because God was able to convey those ideas to "simple" humans.

This once again leaves the Christian with the straightforward reading of the text. Such a conclusion is, however, problematic if indeed God used billions of years of Darwinian evolution to create the universe. If the latter account of creation is true, then the account presented in Genesis 1 is false. This brings the "deception" accusation full circle. Either the account found in Genesis 1 is historically accurate, or God is lying to us. This time, however, the deception is found not in vague natural science, but in easily understood propositional truths.

Chapter 10

Final Thoughts

This book is only one facet of a very complex, yet fascinating area of Christian creationist apologetics. I hope that as you have read through this book's pages, your heart has been stirred to trust the Word of God *above* that of your senses and reason. That is not to diminish the latter, it is just that our senses and/or reason cannot be *completely* trusted. All too often we choose what seems wise or best, based solely on a feeling or in conjunction with what our eyes or mind tells us is true. To reason biblically means to reject our feelings, our experience, and even what we see, if any of these communicative avenues *clearly* contradict God's revealed Word. This is so relevant today, not just in YEC circles, but in the church at large, where Christians, longing for "something more" make perilous doctrinal decisions. We should take God at His Word when the Scriptures say: "Trust in the Lord with all your heart, and do not lean on your own understanding" (Proverbs 3:5 ESV).

To my YEC critics

No doubt, I will have many critics and detractors on both sides of the origins fence. Many YEC, for example, will find little comfort in "giving up" scientific ground to the "evolutionists." Yet as this book clearly shows, YEC must foster a spirit of humility when applying the scientific method to certain parts of the created order. Many aspects of creation can and should be interpreted only in a "naturalistic"[1] sense. Given this, many might think this book *overly* naturalistic. After all, if the processes God currently uses to sustain the universe were the same ones He used to construct it in the first place, we are left with nothing more than 18th century Deism. This perspective, however, misunderstands God's power. Clicking His fingers and making wine from water is not the only way God creates. The slow and gradual growth of mountains all over the world also attests to God's creative power. Now, it may be that some interpret this growth in purely materialistic terms, but as Poythress (2019, p. 32) suggests, this is because they refuse to see it any other way:

> *Yes, people can tell themselves the tale that the regularities found by scientists are part of an impersonal mechanism rather than an expression and display of the faithfulness of God in his rule over all. But the tale is false. And it can be shown to be false, because the regularities themselves are rational and language-like, testifying to the personal nature of the God who specifies them.*

God is working all things according to His will. This includes mountains, rocks, dirt, dust, flowers, butterflies, sharks, bees, flies, planets, suns, stars, solar systems, atoms, protons, neutrons, electrons,

---

1 Again, for fear of being misquoted, I want to emphasize what I mean by naturalistic. I use the term to emphasize God's providential rhythms of regularity in nature. God uses supernatural, non-rhythmic creative powers, but, according to Scripture, He also uses natural, rhythmic sustaining powers. This is what I mean by "naturalistic." Do not confuse naturalistic with naturalism. The latter term denies supernatural events and divine superintendence in nature.

fermions, quarks, and leptons. All these things are working according to God's command through laws that *God* has set in place under the scrutiny of *God's* providence. Christians need to stop thinking that God gets the glory only for supernatural acts of fiat creation. He gets the glory for every breath we take, every drop of blood that pumps around our bodies, and every action potential that crosses our neurons.

It may be that some YEC will not like the overly *philosophical* nature of this book, perhaps even seeing an unscriptural and/or liberal focus that denies the authority of the Bible. Nothing, of course, could be further from the truth. Unfortunately, there are many YEC that equate certain YEC positions with the doctrines of Scripture. I think I have been very careful to stay within the bounds of what is *biblical.* I understand, however, that there must be some "absolutes." Exactly what those absolutes are is a matter of discussion, but I think the list put together by Warren Johns (2016) is incredibly helpful:

- The days of Creation are literal historical days, not day-ages, analogical days, revelatory days, literary days, mythological days, theological days, or any other type of metaphorical day.
- The Flood was a universal, catastrophic event that has left its mark upon the geological record.
- The genealogies and other historical records stretching from Adam to the present suggest a short-age time frame. The short time frame includes the entire universe.
- The mega-evolution of Darwinism is not accepted as a paradigm for earth history, but on the level of micro-evolution significant changes are allowed within the original created "kinds."
- The Bible from Genesis to Revelation is inerrant in its essence and is accurate in the areas of history and science as well as in theology.

Where *can* the YEC biblically deviate from this list? This is important because there are many non-absolute avenues of research that continue to support a biblical perspective. YEC must be allowed to explore those avenues without being accused of heresy. I propose keeping Johns' (2016) list, but further evaluating some of the points in that list. Here are my thoughts:

- The days of Creation are literal historical days, not day-ages, analogical days, revelatory days, literary days, mythological days, theological days, or any other type of metaphorical day.

Although all YEC must acknowledge a literal six-day Creation Week, there is room for conjecture as to what *occurred* during that time. Any perspective is valid, as long as it does not contradict Scripture.

- The Flood was a universal, catastrophic event that has left its mark upon the geological record.

The Flood of Noah was a real, world-wide watery catastrophe. As such, geological evidence should exist that supports this catastrophe. Many YEC, however, have overstepped the line here by proposing that almost *all* the world's sedimentary layers were deposited in the Flood. To think otherwise is sometimes viewed sacrilegiously. Now, it *may* be that nearly all the Earth's geological layers were in fact deposited during Noah's Flood, but there is nothing in Scripture that *demands* this. For all we know, the entire Flood record could extend to a single geological package of strata that represents a single geological period. All that is required is *significant* geological evidence, not *all* geological evidence.

- The genealogies and other historical records stretching from Adam to the present suggest a short-age time frame. The short time frame includes the entire universe.

The genealogies are a strong internal evidence as to the youth of the biosphere, including the origin of man. Since Adam is tied to the six literal creation days, then the age of the universe is unequivocally in the thousands of years. There may, however, be some gaps in those genealogies. The gaps cannot be too large, but they could be large enough to push back creation to perhaps as many as 10,000 years ago.

- The mega-evolution of Darwinism is not accepted as a paradigm for earth history, but on the level of micro-evolution significant changes are allowed within the original created "kinds."

I like that Johns (2016) talks about "significant" evolutionary changes. Christians have been too conservative in this area for far too long. This is most likely a holdover from the incorrect creationist position of the 17th through 19th centuries, which held to the fixity of each species (i.e., the belief that species did not evolve *at all*). More recently, YEC have acknowledged this error and have moved into some interesting, if not controversial, areas regarding biological evolution. No one is saying that man evolved from ape-like ancestors or that dogs evolved from cats, but some are suggesting that pretty significant change has occurred in some organisms. This has brought flak from more "conservative" YEC who have accused these people of siding with TE. Again, however, nothing could be further from the truth. YEC should be willing to concede ground here. There is much room in the text that allows for lots of speculation on this issue.

- The Bible from Genesis to Revelation is inerrant in its essence and is accurate in the areas of history and science as well as in theology.

There is no need to elucidate on the last point. With this, all Christians should agree.

To my OEC and TE critics

From the OEC and TE communities, I am sure I will also get an ear-bashing. I will most likely be accused of simplistically brushing YEC problems under the proverbial rug. After all, if we can fit almost the entire cosmos, bar a veneer of sediment at Earth's surface, into the Creation Week, and if we hold that all of these created objects have an apparent age, then are we not just "having our cake and eating it too"?

In other words, on the one hand we get to believe that the universe is young, but on the other we conveniently get to disregard the scientific evidence that indicates it is old. I want to spend some time fleshing out this objection.

First, it is not convenient at all. It is true that most of the created universe falls into this category, but this is by God's design, not that of the YEC.

Second, it makes sense that the only "normal" cosmological history belongs to the time/historical context related to God's crowning creation—man. Extrapolating time back billions of years is a common secular notion concerned with diminishing the importance of man. This is very similar to the cosmological perspective that contradicts God's sequential order in creation. By doing this, the secular, scientific community has successfully relegated man, God's crowning achievement, to an accidental grouping of carbon atoms that have existed on an unimportant little blue dot for the briefest of moments.

Third, God has provided multiple examples of SFP. This is not a "straw-grasping" exercise. It is firmly grounded on the analogy of Scripture and a biblically *explicit* doctrine.

Finally, there are still many other scientific problems in modern young-Earth creationism that this book does not address. As an example, consider the problem of Deep Time. Although SFP theory accounts for accelerated rates of isotope decay in rocks associated with the Creation Week, it does not account for rates of accelerated decay in rocks deposited *after* Creation Week. Yet modern isotope dating techniques clearly show that Flood rocks and post-Flood rocks have isotope ratios that reflect millions of years' worth of decay as interpreted by the secular scientific community. This problem cannot be solved by appealing to SFP theory, since these events occurred in "real time" and so must be addressed using the normal rules of science.

To the scientist, pastor, Bible teacher, and Bible scholar

It is upsetting to see many pastors and Bible teachers capitulating to old-Earth and/or theistic evolutionary ideas. Given misconceptions that come from all three camps, including young-Earth creationism, I sincerely understand why this is occurring.

This is not to say that those who espouse such views are non-Christians. I believe that many are sincere believers grappling with difficult questions. Having said this, I do believe that theistic evolution *in particular* is an incredibly dangerous doctrine that has a façade of biblical orthodoxy. Theistic evolution denies many key tenants of The Faith including, but not limited to, biblical inerrancy, a literal Adam, a literal Fall and original sin.

In comparison, many committed to old-Earth creationism tend to adhere to these doctrines. As such, I am a lot more sympathetic to those who hold to an old-Earth position and am actually very thankful for the concerted effort many are making as they seek to undermine theistic evolution (Moreland et al. 2017).

That said, I still want to challenge scientists, pastors, teachers, and Bible scholars who teach old-Earth creationism, but especially those teaching theistic evolution: Do you ever stop and think about the influence you are having on your audience? As I wrote this book, I consistently met the Lord in prayer, at times fearfully, pleading with Him to keep my words *true*. Although I might be ignorant on some matters of science, and perhaps even misguided, I know that I am not mistaken about the text of Scripture. I know that the Bible teaches man was specially created by God on Day 6 of Creation Week. I know that the Bible teaches man fell into sin, and I know that this spiritual malady is the cause of this world's troubles. This is just what *the text* says.

I am comfortable publishing a book that affirms this and my conscience is at ease. My question to scientists, pastors, teachers and Bible scholars who teach and publish is: Are you comfortable teaching and publishing *your* ideas? "Not many of you should become teachers, my brothers, for you know that we who teach will be judged with greater strictness" (James 3:1 ESV).

What about the life sciences?

This book has addressed YEC issues only as they relate to the *physical* sciences. What about issues related to the *life* sciences? The similarity of biological systems and organisms is thought to be one of the most powerful arguments in favor of Darwinian evolution. In response to that, creationists of all persuasions will typically say something like: "Common design indicates a common designer." And that is okay, this is a good defense.

The problem is that it does not really explain the *how* of this design. The secular scientific community, however, has a robust model explaining the similarity and diversity of biological systems—universal common ancestry through natural selection over geologic time. Everything has similarity because everything is *related*.

I think, however, that some of the concepts applied in this book could be helpful in developing a robust *creationist* model for biological relatedness. Since all organisms alive today came into existence at some point *after* Creation Week, unlike much in the inorganic realm, there is no way of applying SFP theory to biological entities. The importance of *process*, however, applied to *conceptual* solar systems and galaxies may provide a working model explaining biological relatedness.

Some of the following was discussed in previous chapters but will be rehashed here for the purpose of clarity. We know that the functioning of Adam's biology, right down to the intricacies of the living cell, must have been planned out to the finest detail before Adam was created. We know this because the whole plan of salvation was set in motion before a single atom had sprung into existence—a plan that involved the second person of the trinity *taking on flesh*. God could not have planned all the redemptive aspects related to salvation without also planning Jesus' biological ones. By default, then, Adam's biology must also have been planned from eternity past.

We know that biological structures can change by manipulating genes. Darwinian evolution wants to take the credit for this manipulation by giving the power of genetic change to nature—hence *natural* selection. The problem from a materialistic perspective is that the secular scientific community has only scant clues as to *how* the information responsible for generating natural selection came about (Moreland et al. 2017).

Enter the creationist. Christians should be giving God the glory for the amazing changes that can occur within biological organisms. God has built an incredibly complex system into organisms that allows them to change as they adapt to their environments.

YEC Randy Guliuzza and Phil Gaskill (2018) have developed an idea that puts the spotlight back where it belongs—on the Creator. In their paper, they discuss the importance of *pre-programmed* internal genetic mechanisms. In the secular model, natural selection must act on mutations that have *randomly* found their way into the genome. In Guliuzza and Gaskill's model, most of the information was already there *from the beginning*—hence, pre-programmed.

Okay, so how does all this help when it comes to relatedness and Darwinian evolution? Well, what if similar processes that currently diversify organisms today were at work in the mind of God prior to Creation Week for the purpose of designing them? What if God took our conceptual Adam and "worked on"[2] Adam's genetic information for the purpose of producing his closest biological cousin—monkeys?

Once God has a conceptual monkey, He can then "work on" this genetic material and create *their* closest biological cousin—lemurs. This process could continue until God has an entire conceptual biosphere in His mind, all ready to bring into the time/historical context of Creation Week in any manner He sees fit. Once these organisms are brought into the time/historical context of Creation Week, they continue to change as per the Creator's genetic instructions. These instructions do have boundaries, and so each group of organisms can change only to a limited extent. Yet the way in which these organisms change over time is related to *the same mechanisms that designed them in the first place*. This is in line with God's *anticipating* a natural universe and natural laws that would function according to natural processes even before any of these processes were brought into existence.

---

2 "Worked on" is purposefully vague.

This is not accelerated evolution in the mind of God. Secular evolutionary models want the information necessary for change to come into existence *randomly over time*. Yet if Guliuzza and Gaskill (2018) are correct, then most of this information has been there from the very beginning, and even prior to the beginning when it was used to design the organism in the first place. This model acknowledges relatedness *without* the need for "common ancestry," with the latter term inherently invoking ancestry, and thus demanding an evolutionary scenario that plays out over time. This means that yes, man is more closely "related" to an ape than he is to a lemur, but this relatedness *did not come into existence during millions of years of evolution*; it took place *in the mind of God*, prior to creation, with each organism being placed into its future ecological/geographic environment on specific days of the Creation Week and then allowed to change within pre-programmed parameters over time.

Importantly, this model of biological relatedness differs from Aristotle's *scala naturae* (Merged into the Great Chain of Being of the medieval period). The Great Chain of Being, as espoused by many early Christians, presented a tiered hierarchy of organisms with God at the top and lower life forms at the bottom. In Aristotle's *scala naturae*, and later in the Great Chain of Being, these organisms were "fixed" and unable to change. This idea was later developed and unfortunately incorporated into the scientific milieu of early biology. It became known as the "fixity of species."

Although Christians should reject many aspects of the Great Chain of Being, there may yet be much to commend a hierarchical view of creation (Wood and Murray 2003, p. 30). With a little work, I believe that the ideas proposed above, in conjunction with the work of others (Wood and Murray 2003), may yet deliver a simpler and more robust model of biological relatedness.

I get it

If you are a believer and struggling with the science, I get it. But here is my advice to you.

As you have read this book, no doubt you have come to see that I put a lot more stock in the validity of secular scientific interpretations than many overly skeptical Christians do. That is because most scientists are not out to wantonly lead people astray. They are just normal people who love science and want to contribute to humanity's love affair with knowledge for the purpose of furthering many aspects of human endeavor. This is a good thing. As such, Christians should never be afraid of science. Having said that, I believe there are important facets of science that Christians need to keep in mind.

First, most scientists do research in very *specific* areas—studying the appendages of a particular kind of bug, doing experiments on the hydrogeological aspects of moving water, comparing the courtship behavior of different birds. As such, all scientists *heavily* rely on the expertise of *other* scientists for knowledge *outside* of their specific areas of research. As a non-scientist, I remember thinking that scientists know absolutely everything. If a scientist said it, it was true. *As* a scientist, I have come to appreciate that I was greatly mistaken. Consider my own field: Upper Cambrian stromatolites of North America (Coulson and Brand 2016). I can tell you all you want to know about these structures. I can tell you about their morphologies, the likely paleo-environment in which they grew, what they are composed of, as well as a number of other very interesting facts, but I could not tell you too much about the structural aspects of the Canadian Shield in Canada. I rely heavily on the work of other geologists for that information. I rely even more on the work of biologists and astronomers to tell me something about the reality of life and the nature of the universe. Biochemist and ID advocate

Douglas Axe (Axe 2017) admits that this problem even exists within the same field:

> *As a biochemistry professor at the University of Southampton in the UK, Fox is an established expert on how various molecules bind to DNA. Having done no research on that subject, I'm obviously in no position to critique his work. Likewise, having done no work on protein evolution, he is really in no position to critique my work (Axe 2017, p. 84).*

Axe goes on to report that Fox, an evolutionist, made a public claim about abiogenesis that impinged upon Axe's specialization and not his own. What is more, Fox made an error about that claim. If I had read Fox's claim *first*, I, having no knowledge of biochemistry myself, would more than likely have accepted his claim on the assumption that he was an expert on that topic when, in fact, he was not. All of us must be very humble with how much we know and how much we *think* others know.

Second, although I do not believe in a scientific conspiracy out to wantonly lead Christians astray, I do believe that all Christians are in a spiritual battle: "For the weapons of our warfare are not of the flesh but have divine power to destroy strongholds. We destroy arguments and every lofty opinion raised against the knowledge of God, and take every thought captive to obey Christ" (2 Corinthians 10: 4 – 5 ESV). The Bible clearly teaches that the entire world is under the power of a deceptive, spiritual force (2 Corinthians 4:4). The apostle Peter tells us that this unseen power, Satan, is out to spiritually "devour" those unwaveringly committed to God's standard of spiritual Truth (1 Peter 5:8). Skeptics are incredulous that anyone could believe such "nonsense," but Christians know that spiritual warfare is a reality and that without God's Spirit, there can exist only spiritual blindness. Scientists are not immune from Satan's influence. Taken together with the sinful, fallen and selfish heart of man (Jeremiah 17:9), Christians should be very cautious about large-scale scientific theories that seek to describe reality.

Third, philosophically speaking, everybody should know that science is limited in its ability to convey accurate descriptions of reality. Scientific methodology is a great tool and has much to commend it, but it is not comprehensive (Moreland 2019). Many people, scientists included, do not know this. Thomas Kuhn, Karl Popper, J.P. Moreland and many other philosophers of science have persuasively demonstrated that science rests on the philosophical framework that *itself* defines the nature of science. Since science is not the epistemological top dog, the average person in the street should not automatically assume that large-scale descriptions of reality are necessarily scientific.

Fourth, God's supernatural presence is purposefully detached from His creation. As Wise (2002, p. 12) puts it: "So God has placed enough flexibility in His creation to permit us not only to reject truth but to create, accept, hold to, and act out non-truths." Given this, it makes sense that there exist alternate interpretations for natural phenomena. As discussed in Chapter 8, this is not the act of a mean, deceitful or vindictive God, it is an act of grace towards rebellious and hostile subjects who reject God as He has revealed Himself in Scripture. A *strong* sense of God's existence in nature coupled with hostile subjects will result only in swift judgement. One day, this exact situation will be played out:

> *The rest of mankind, who were not killed by these [supernatural] plagues,* ***did not repent*** *of the works of their hands nor give up*

> *worshiping demons and idols of gold and silver and bronze and stone and wood, which cannot see or hear or walk,* ***nor did they repent*** *of their murders or their sorceries or their sexual immorality or their thefts (Revelation 9:20 – 21 ESV; emphasis mine).*

Yet there is still hope: "The Lord is not slow to fulfill his promise as some count slowness, but is patient toward you, not wishing that any should perish, but that all should reach repentance" (2 Peter 3:9 ESV). The dispensation that presently exists between Christ's first and second coming is a *time of grace*. It is a dispensation that exists for the sole purpose of allowing the rebels to repent and bow their knees before God Almighty. Do not confuse this grace and the ambiguity found in nature with the ultimate Truth of the universe as it is revealed in Scripture.

Finally, young-Earth creationism is a recent movement, on the order of about 50 years. What we need is young, energetic, open-minded scientists to join the ranks and start thinking outside the box for the purpose of defending a cohesive and scientifically sustainable model of young-Earth creationism. We also need money for research. There are hundreds of projects out there for inquisitive young minds to get their scientific teeth into. Given the people and the money, I am sure much more progress will be made, but these are very, very early days.

Given all the above, I encourage the troubled believer to keep the faith. Do not capitulate to old-Earth, or even worse, theistic evolutionary ideas. Become a fool for Christ and get involved.

If you are reading this book and are *not* a believer, more than likely it is because you are interested in Christianity but find some aspects of this belief system scientifically irrational. Even today, after 27 years in the faith, I still sometimes feel this way. Hang in there.

Pray for God to open your eyes and reveal Himself to you. Above all, realize that the Christian life is most definitively a life of faith (Hebrews 11:6). And this makes sense, especially given all that has been said above.

The universe is vast, knowledge is boundless. Your experience with the universe most likely spans a few decades, perhaps as many as nine. Yet compare that experience with all that has transpired on Earth and within the universe over all time.

Faith is truly the order of the day. Trust in Christ and receive his mercy.

Appendix A

Different Dating Methods

The following is a list of dating methods used in the secular, as well as YEC, scientific communities. I have included only those examples that *constrain* an "age" either one way or the other. For example, the recession of the moon and the faint sun paradox do not furnish evidence supporting either a 10,000-year-old *or* a multi-billion-year-old solar system. These two examples are used by YEC *only* because they constrain a *4.6 billion-year-old* solar system.

The following examples have been chosen because they *constrain* the time *one way or the other*—either a universe *no older* than a few tens of thousands of years or one that is at *least* multi-millions of years old or older.

Importantly, the following examples represent systems that were set in motion during Creation Week. Many YEC do not distinguish the six days of creation from the 6,000 years that have passed *since* creation. Processes that left a record in *post*-Creation Week sediments, minerals or rocks, however, can in no way be used to extrapolate "ages" for processes that were set in motion *during* Creation Week. To say, "the Earth is young" using rates at which sediment fills ocean basins, therefore, is not precise. What is young is the sediment, *not the Earth* (i.e., the core, mantle and/or crust).

Since SFP theory strongly suggests that all rates, everywhere were the same, *relative to each other*, during Creation Week as they are in the present, and since it also strongly suggests that these rates were accelerated billions of times, then Creation Week systems should be stamped with vast amounts of process, interpreted today by the secular scientific community in terms of billions of years.

Processes indicating the general "youth" of creation (less than tens of thousands of years)

Comets: **a solar system less than 10,000 years old.**

> Comets are chiefly made of ice. As comets orbit the sun, they lose some of their mass to the solar winds. As such, comets should exist only for thousands of years before completely burning up.
> https://answersingenesis.org/evidence-for-creation/the-10-best-evidences-from-science-that-confirm-a-young-earth/

Earth's decaying magnetic field: **an Earth less than 20,000 years old.**

> The Earth's magnetic field is decaying at an exponential rate. Extrapolating this rate backwards would furnish the Earth with an "age" of less than about 20,000 years before the field totally collapsed.
> https://answersingenesis.org/evidence-for-creation/the-10-best-evidences-from-science-that-confirm-a-young-earth/

Moon dust: **no longer used by YEC due to recent scientific discoveries (see Faulkner 2018b).**

Processes indicating the general "antiquity" of creation (more than a few million years)

Stellar evolution: **a universe that is at least millions to tens of millions of years old.**

The main process responsible for producing photons of light is the fusion of hydrogen. Yet millions to hundreds of millions of years are required for a star to reach a useful, light-emanating age.
https://www.space.com/24054-how-old-is-the-universe.html

Saturn's rings: **a solar system that is at least 10 million years old.**

The rate at which dust settles on the rings of Saturn has recently been established. Extrapolating this rate backwards suggests that the rings must be at least 10 million years old, but perhaps as old as 100 million years.
https://www.futurity.org/saturn-rings-age-1964302/

Radiohalos: **an Earth that is at least 100 million years old.**

One hundred million years' worth of radioisotope decay is required to form a single, dark uranium halo. Halos form when minerals containing uranium are etched by the decay of alpha particles. Note, this process is purely mechanical due to the movement of small nuclei through a solid. Although there are issues with radioisotope decay methods, this purely mechanical mechanism leaves little room for error as to the millions of years' worth of decay at today's rates.
https://answersingenesis.org/age-of-the-earth/radiohalos-solving-the-mystery-of-the-missing-bullets/

Spiral galaxies: **a universe that is at least hundreds of millions to billions of years old.**

Spiral galaxies have long "arms" containing hundreds of millions of stars that extend out away from the central region of the galaxy. The closer the stars are to the central region of the galaxy, the faster the rate at which the arms containing those stars wind up. This means the galaxy is eventually going to be wound up very tight. Extrapolating backwards the current rate at which these galaxies are winding up furnishes an "age" in the hundreds of millions of years to billions of years.
https://www.scientificamerican.com/article/what-process-creates-and/

Colliding galaxies: **a universe that is at least billions of years old.**

Billions of years are required to "untangle" entwined, colliding galaxies using the current rate at which they are merging.
https://www.technology.org/2019/03/14/milky-way-and-andromeda-galaxies-will-look-like-this-after-collision-billions-of-years-from-now/

Expanding universe: **a universe that is at least billions of years old.**

Billions of years are required to account for the expansion of the universe. Now, although this expansion is debated, there is still enough movement to suggest billions of years at the current rates at which galaxies move apart.
https://www.space.com/24054-how-old-is-the-universe.html

Earth's core, mantle and crust: **an Earth that is at least a few billions of years old.**

Fractionation, partial melting and gravitational differentiation are all processes that currently produce continental crust at exceedingly slow rates. Extrapolating these rates backwards in time to account for the 7 billion cubic kilometers of granitic rock requires billions of years (Baumgardner 2000).[1]

Exploding supernovae: **a universe that is at least 10 billion years old.**

The "age" of some supernova has been calculated at over 10 billion years using the rate at which the supernova's dust cloud expands. Extrapolating this rate backwards requires very large time frames.
https://www.southampton.ac.uk/news/2018/02/oldest-supernova-discovered.page

Star light: **a universe that is at least 13 billion years old.**

Multiple billions of years are required to account for the time it took for the light from distant stars to reach Earth.
https://www.forbes.com/sites/startswithabang/2018/02/23/if-the-universe-is-13-8-billion-years-old-how-can-we-see-46-billion-light-years-away/#30ee2cae1303

1 Baumgardner strongly infers this but does not actually state it.

Appendix B

Explaining the Concept Simplistically

Explanations of the scientific processes outlined in certain chapters are here fleshed out at a simplistic level of understanding. Please be aware that precision and accuracy have been sacrificed for the purpose of clarity.

Polonium halos from Chapter 5

When the element polonium decays, small pieces of it shoot away from its center at high speeds in all directions. As each piece moves through solid rock, it cuts a path into the rock. This would be like shooting a bullet into ice: The bullet would cut through the ice and leave a path that you could see. Since these pieces shoot away from the polonium in all directions, all the little paths together etch a perfect sphere into the rock. Hence a polonium halo.

Now imagine that God creates hot liquid magma and then puts a whole bunch of polonium in the magma. Since polonium decays away in just a few minutes, how much would be left in the magma after a few hours? None! This means that when the magma finally hardens into rock, no polonium will be left to get "stuck" in the rock.

Now let us imagine another scenario. This time God makes the magma and puts the polonium in just like before, only this time God supernaturally causes the magma to cool and turn into rock in *less than a minute*. Will any of the polonium get "stuck" in the rock? Yes. This is because it takes *more* than a minute for it to all decay away. Just because the polonium gets stuck in the rock, however, does not stop it decaying. Since the polonium decays by shooting little pieces out from its center in all directions, it will etch a halo into the rock. It could not do that before because it decayed in a liquid, but because rock is a solid it gets damaged (the halo is the damage).

Some creationists think this means that the rocks were created instantly. If the magma was left to cool over millions of years, how much polonium would be left in the rock? None! This means that when the magma finally does turn into rock, there will be *no* polonium left to get stuck in the rock, which would also mean there wouldn't be any polonium halos. The *presence* of halos means that the magma cooled supernaturally in less than a few minutes.

Most creationists do not like this idea anymore because a solution was found that solved the apparent problem. As the element uranium decays, it turns into another element called radon, which is a gas. When the radon decays, it turns into polonium. It turns out that the polonium sort of "lives" inside the uranium. Think of uranium, radon and polonium like nesting dolls. As you peel away one layer, there is another one beneath it. The polonium, however, must wait a very long time before the uranium and the radon are peeled away to reveal the polonium. That is because it takes millions of years for the uranium to decay, unlike the polonium that just takes a few minutes.

Now imagine that God creates a magma and puts *uranium* into the magma. If it took one million years for the magma to turn into rock, would uranium get stuck in the rock? Yes, because it takes *more* than a million years for it to decay away. Since the polonium lives inside the uranium, it can stay "preserved" for the entire million years. Once the rock has cooled and hardened, the polonium escapes from the uranium when it finally decays after millions of years. This means the polonium is free to decay in just a few minutes, leaving a halo in the rock. In this scenario, the magma did not have to cool and harden in just a few minutes since the polonium was sort of "waiting" inside the Uranium.

Earth's decaying magnetic field from Chapter 5

Our planet has a big magnet inside of it. The top of the planet is a north pole and the bottom of the planet is a south pole. Like any kind of magnet, there is a field associated with it, and this field has a certain strength. Scientists have been measuring the Earth's magnetic field strength for many years, and it turns out that it is getting weaker and weaker. Eventually, the field will get to near zero strength.

This is not good because when the field gets to near zero strength, it can no longer protect the planet from all sorts of rays that come from the sun. Since the field is getting progressively weaker, scientists have been able to figure out how strong the field was in the past (Figure 1B). It turns out that if you go back more than about 10,000 years, the field was so strong that life would not have been possible. What we need on Earth is a field that is not *too* weak but also *not* too strong, otherwise in both scenarios, life would not be possible. Creationists have suggested that this proves the Earth is no older than 10,000 years. If it was older than that, then life would not be able to exist because the magnetic field was too strong.

The secular scientific community soon discovered that, indeed, this evidence conflicted with an Earth that is supposed to be billions of years old. In order to solve this problem, scientists proposed that the field is indeed getting weaker, but the reason for this is that the magnet inside the Earth "flips." Our present north pole is getting weaker and weaker, and will eventually reach a near zero strength, but when it does, the north pole will flip and become the south pole, and the south pole will flip and become the north pole. When this happens, the field strength will start to get stronger again. Eventually, in tens of thousands of years, the field will again go from getting stronger and stronger to getting weaker and weaker, just like it is now. Then the whole process will start again. From a secular, scientific perspective, the Earth's magnet field is about ready to flip again. This is why it is currently so weak.

This idea has some good evidence to support it. The best evidence is magnetic striping on the ocean floors. As magma erupts on the sea floor, tiny bits of iron that are in the magma act like little compass needles and point in the direction of the Earth's magnetic field. When the magma cools and becomes rock, those little iron needles in the rock get frozen in place. Geologists can look at the ocean floor and work out what direction the Earth's magnetic field was in the past just by looking at the direction of those little iron needles. It turns out that Earth's magnetic field has indeed flipped hundreds of times in the past (Figure 1A).

There is still a problem with the secular theory, however. Although everyone agrees that the poles have flipped, not everyone agrees about how the magnetic field gets stronger and stronger after a flip. Scientists have yet to come up with a model that fully answers this question. The model that secular scientists appeal to is called the Dynamo Effect. We know that there is hot, liquid iron in the Earth's core. This iron liquid has lots and lots of electrons in it. Scientists suspect that heat within the core causes these electrified fluids to move, just like when you heat up water on the stove. Moving electrons produce electricity. If you move electrons through a wire using a battery, for example, you get a light bulb to glow. What's interesting about these moving electrons is that not only do they get the light globe to glow, they also produce a magnetic field around the wire! That is right: Put a compass next to a flashlight, and it might just move your compass needle! Scientists say this is what happens in the Earth's core. As these electrified fluids of iron move, they produce a magnetic field that gets stronger and stronger.

The original creationist model that described the decay of the magnetic field looked like figure 1B. It was thought that God created the Earth with a north pole and that this field slowly decayed over time but *never* flipped. In the 1960s, however, scientists found that, indeed, the magnetic field had flipped hundreds of times in the past (Figure 1A). When this new evidence came out, creationists incorporated these flips into the original model from figure 1B. This new model is what we see in figure 1C. Notice that it looks like figure 1B but has squiggles in the middle. Some creationists now suggest that Earth's magnetic field flipped multiple times in the past because of Noah's Flood. As large parts of the ocean floor got sucked down into the Earth's mantle, they eventually "hit" the core, causing the electrified iron fluids to move a lot. This movement caused the fluids to produce magnetic field reversals (flips) that continued for perhaps hundreds of years after the Flood. According to many YEC, these flips have since stopped and the field will not reverse ever again (Figure 1B).

Earth's decaying magnetic field from Chapter 7

In Chapter 7, I proposed that the magnetic field may, indeed, have been "flipping" since Creation Week, and that the Flood of Noah simply accentuated this flipping. This is contra Humphreys (1984), who says the flipping occurred only during and shortly after Noah's Flood. I do not, however, think that the secular time frames of tens of thousands and hundreds of thousands of years for each reversal are correct.

During these lengthy times, the field strength is moderate to very high, which means the field protects the Earth from dangerous rays emanating from the sun. According to secular scientists, however, there is a short amount of time involved in the actual "flip" of the field (from north to south or vice versa), on the order of about 7,000 years (Clement 2004). But secular scientists realize that this presents a problem (Wei et al. 2004). Seven thousand years is too long for the magnetic field strength to be close to zero. Such a long period of time without a magnetic field means that the Earth's oxygen would get sucked into outer space. This would lead to a mass extinction on our planet. Yet if the magnetic field has flipped hundreds of times, then the fossil record should show evidence of hundreds of mass extinctions. But it does not! There are only a handful.

Since the Earth is only about 10,000 years old, the hundreds of flips that have occurred must have lasted for very *short* time frames relative to the secular model that believes each reversal lasted for tens of thousands to hundreds of thousands of years. One of the problems in trying to figure out the "normal" length of a field reversal is that most of the ones recorded in the ocean floors resulted from rapid reversals (days, weeks, months?) that occurred during the Flood of Noah. Nevertheless, short field reversals on the order of just a few hundred years or perhaps a few thousand years best fit the fossil record and the lack of mass extinction events.

References

Augustine of Hippo. 354 – 430. *Of the Merits and Remission of Sins and of the Baptism of Infants*. Book 1, chapter 2.

Allwood, C.A., Walter, M.R., Burch, I.W., and Kamber, B.S. 2007. "3.43 billion-year-old Stromatolite Reef from the Pilbara Craton of Western Australia. Ecosystem-scale Insights to Early Life on Earth." *Precambrian Research* 158: 198 – 227.

Axe, D.D. 2017. "Three Good Reasons for People of Faith to Reject Darwinist's Explanation of Life." In *Theistic Evolution: A Scientific, Philosophical, and Theological Critique*, edited by J.P Moreland, S.C Meyer, C. Shaw, A.K Gauger and W. Grudem, 83 – 104. Wheaton, Illinois. Crossway.

Berkhof, L. 1938. *Systematic Theology*. Pdf version available at: http://www.apuritansmind.com/wp-content/uploads/2018/03/Systematic-Theology-by-Louis-Berkhof.pdf

Barnes, T.G. 1982. "Young Age for the Earth and the Moon." Accessed February 23, 2019. https://www.icr.org/article/young-age-for-moon-earth/

Baumgardner, J.R. 2000. "Distribution of Radioactive Isotopes in the Earth." In *Radioisotopes and the Age of the Earth*, edited by L. Vardiman, A.A. Snelling and E.F Chaffin, 49 – 94. Dallas, Texas: Institute for Creation Research, and Chino Valley, Creation Research Society.

Barrick, W.D., forthcoming. *Genesis 1 – 11, Evangelical Exegetical Commentary*. Bellingham, WA: Faithlife/Lexham Press

Bradie, M., and Duncan, C. 1997. "The Evolution of the Concepts of Space and Time, Third Preliminary Version, Chapter 4, Plato's Universe." Accessed January 30, 2020. http://physics.bgsu.edu/~gcd/Spacetime4.html

Brand, L., and Chadwick, A. 2016. *Faith, Reason, and Earth History: A Paradigm of Earth and Biological Origins by Intelligent Design, 3rd edition*. Michigan: Andrews University Press.

Brand, L. 1979. "Field and Laboratory Studies on the Coconino Sandstone (Permian) Vertebrate Footprints and Their Paleo-ecological Implications." *Palaeogeography, Palaeoclimatology, Palaeoecology* 28: 25 – 38.

C. B. 1853. *Geology in its Relation to Revealed Religion*. Dublin, Ireland: Gerald Bellew.

Clarke A. 1810 – 1826. *Adam Clarke's Commentary on the Bible*. https://www.studylight.org/commentary/hebrews/9-23.html

Clarey, T. 2018. "Stromatolites: Evidence of Pre-Flood Hydrology." *Acts & Facts* 47, no 5.

Clement, B. 2004. "Dependence of the Duration of Geomagnetic Polarity Reversals on Site Latitude." *Nature* 428: 637 – 640.

Coulson, K.P. 2018. "Global Deposits of in situ Upper Cambrian Microbialites—Implications for a Cohesive Model of Origins." In *Proceedings of the Eighth International Conference on Creationism*, edited by J.H. Whitmore, 373 – 388. Pittsburgh, Pennsylvania: Creation Science Fellowship.

Coulson, K.P., and Brand, L.R. 2016. "Lithistid Sponge-microbial Reef-building Communities Construct Laminated, Upper Cambrian (Furongian) 'Stromatolites.'" *Palaios* 31, no. 7: 358 – 370.

Coulson, K.P., Brand, L.R., and Chadwick, A.V. 2016. "Microbialite Elongation by Means of Coalescence: An Example from the Upper Cambrian of Utah." *Facies* 62. https://doi.org/10.1007/s10347-016-0469-5

Deem, R. 2013. "Appearance of Age. A Young Earth Problem." Accessed March 14, 2019. http://www.godandscience.org/youngearth/appearance.html

DeYoung, D. 2010. "Mature Creation and Seeing Distant Starlight." *Journal of Creation* 24, no. 3: 54 – 59.

Delitzsch, F. 1874 – 1876. *Commentary on the Epistle to the Hebrews*. Edinburgh: T & T. Clark. https://archive.org/details/commentaryonepis02deliuoft/page/222

Dhuime, B., Hawkesworth, C.J., Delavault, H., and Cawood, P.A. 2018. "Rates of Generation and Destruction of the Continental Crust: Implications for Continental Growth." *Philosophical Transactions of the Royal Society* A 376, 20170403: 1 – 11. https://doi.org/10.1098/rsta.2017.0403

Faulkner, D.R. 2018a. "The Case for Cosmological Redshifts." *Answers Research Journal* 11: 31 – 47. www.answersingenesis.org/arj/v11/cosmological_redshifts.pdf

—. 2018b. "The Current State of Creation Astronomy II." In *Proceedings of the Eighth International Conference on Creationism*, edited by J.H. Whitmore, 36 – 45. Pittsburgh, Pennsylvania: Creation Science Fellowship.

—. 2016. *The Created Cosmos: What the Bible Teaches About Astronomy*. Arkansas: New Leaf Publishing Group.

—. 2014. "Interpreting Craters in Terms of the Day 4 Cratering Hypothesis." *Answers Research Journal* 7: 11 – 25.

Fastovsky, D.E., and Weishampel, D.B. 2012. *Dinosaurs: A Concise Natural History (second edition).* New York: Cambridge University Press.

Garner, P.A. 2009. *The New Creationism: Building Scientific Theory on a Biblical Foundation.* United Kingdom: Evangelical Press, EP Books.

Gentry, R.V. 1986. *Creation's Tiny Mystery.* Knoxville, TN: Earth Science Associates.

Gosse, P.H. 1857. *Omphalos: An Attempt to Untie the Geological Knot.* London: John Van Voorst; Paternoster Row.

Greene, J.W. 2012. "A Biblical Case for Old-Earth Creationism." Accessed May 8, 2019. http://www.godandscience.org/youngearth/old_earth_creationism.html

Guliuzza, R.J., and Gaskill, P.B. 2018. "Continuous Environmental Tracking: An Engineering Framework to Understand Adaptation and Diversification." In *Proceedings of the Eighth International Conference on Creationism*, edited by J.H. Whitmore, 158 – 184. Pittsburgh, Pennsylvania: Creation Science Fellowship.

Harold, L. 2017. "Distinguishing Between Apparent Age and Mature Creation." Accessed May 8, 2019. http://thecreationclub.com/distinguishing-between-apparent-age-and-mature-creation/

Hodge, B. 2007. "How Old is the Earth?" Accessed May 8, 2019. https://answersingenesis.org/age-of-the-earth/how-old-is-the-earth/

Humphreys, R.D. 1989. "The Mystery of Earth's Magnetic Field." Accessed February 10, 2020. https://www.icr.org/article/mystery-earths-magnetic-field/

—. 1984. "The Creation of Planetary Magnetic Fields." *Creation Research Society* Quarterly 21, n. 3: 140 – 149.

Huntoon, P.W., and Sears, J.W. 1975. "Bright Angel and Eminence Faults, Eastern Grand Canyon, Arizona." *Geological Society of America Bulletin* 86: 465 – 472.

Johns, W.H. 2016. "Scriptural Geology, Then and Now." Accessed November 5, 2018. https://answersingenesis.org/age-of-the-earth/scriptural-geology-then-and-now/

Jones, D.S. 1989. "Growth Rings and Longevity in Bivalves." Accessed November 5, 2018 http://www.conchologistsofamerica.org/articles/1989/growth-rings-and-longevity-in-bivalves.html

Kuhn, T. S. 1996. *The Structure of Scientific Revolutions.* Chicago: The University of Chicago Press.

Lamoureux, D.O. 2010. "Evolutionary Creation: Moving Beyond the Evolution Versus Creation Debate." *Christian Higher Education* 9, no. 1: 28 – 48

Lisle, J. 2015. *Understanding Genesis: How to Analyze, Interpret, and Defend Scripture.* Green Forest: Arkansas. Master Books.

—. 2011. *Taking Back Astronomy: The Heavens Declare Creation and Science Confirms it.* Green Forest: Arkansas. Master Books.

—. 2009. "The Search for a Cursed Cosmos." Accessed July 11, 2019. https://answersingenesis.org/astronomy/cosmology/the-search-for-a-cursed-cosmos/

Menninga, C. 1988. "Creation, Time and 'Apparent Age.'" Accessed May 8, 2018. https://www.asa3.org/ASA/PSCF/1988/PSCF9-88Menninga.html

Millam, J. 2011. "Coming to Grips with the Early Church Fathers' Perspective on Genesis." Accessed December 2, 2019. http://www.godandscience.org/youngearth/genesis_days_church_fathers.html

Mitchell, T. 2010. "God Created Things to 'Look Old.'" Accessed May 15, 2018 https://answersingenesis.org/creationism/arguments-to-avoid/god-created-things-to-look-old/

Moreland, J.P., Meyer, S.C., Shaw, C., Gauger, A.K., and Grudem, W. eds. 2017. *Theistic Evolution: A Scientific, Philosophical, and Theological Critique.* Wheaton, Illinois. Crossway.

—. 2018. *Scientism and Secularism: Learning to Respond to a Dangerous Ideology.* Wheaton, Illinois. Crossway.

—. 1989. *Christianity and the Nature of Science.* Grand Rapids, Michigan. Baker books.

Moo, D. 1996. *The New International Commentary on the New Testament: The Epistle to the Romans.* Grand Rapids, Michigan/ Cambridge, United Kingdom. William B. Eerdmans Publishing Company.

Morris, H. 1974. *Scientific Creationism.* Arkansas. New Leaf Publishing Group.

Moskowitz, C. 2011. "What's 96 Percent of the Universe Made Of? Astronomers Don't Know." Accessed July 5, 2019. https://www.space.com/11642-dark-matter-dark-energy-4-percent-universe-panek.html

O'Rourke, J., and Stevenson, D. 2016. "Powering Earth's Dynamo with Magnesium Precipitation from the Core." *Nature* 529: 387 – 389. https://doi.org/10.1038/nature16495

Penn, G. 1825. *Comparative Estimate of the Mineral and Mosaical Geologies, Revised edition.* London, United Kingdom. J. Duncan.

Poythress, V.S. 2019. *Interpreting Eden: A Guide to Faithfully Reading and Understanding Genesis 1 – 3.* Wheaton, Illinois. Crossway.

Purdom, G., and Snelling, A.A. 2013. "Survey of Microbial Composition and Mechanisms of Living Stromatolites of the Bahamas and Australia: Developing Criteria to Determine the Biogenicity of Fossil Stromatolites." Accessed January 30, 2020. https://answersingenesis.org/biology/microbiology/survey-of-microbial-composition-and-mechanisms-of-living-stromatolites-of-the-bahamas-and-australia/

Reynolds, J.M. 1994. "Omphalos: A Defense of an Old Argument." In *Proceedings of the Third International Conference on Creationism*, 459 – 466. Pittsburgh, Pennsylvania: Creation Science Fellowship.

Ross, H. 1994. *Creation and Time: A Biblical and Scientific Perspective on the Creation-Date Controversy.* Colorado Springs, Colorado. Navpress.

Rusbult, G. 2003. "Apparent Age, Part I." Accessed May 15, 2019. https://www.asa3.org/ASA/education/origins/aa-cr.htm

—. 2003. "Apparent Age, Part II." Accessed November 20, 2018. https://www.asa3.org/ASA/education/origins/aa2-cr.htm#i

Safati, J. 2015. "The Universe is Finely Tuned for Life." Accessed March 30, 2019. https://creation.com/the-universe-is-finely-tuned-for-life

Sagan, C. 1994. *Pale Blue Dot: A Vision of the Human Future in Space.* New York. Random House Publishing Group.

Schaff P. 1879 – 1890. *Commentary on Hebrews 11:3, Schaff's Popular Commentary on the New Testament.* https://www.studylight.org/commentaries/scn/hebrews-11.html

Scientific American. 2004. "Mature Galaxies in Young Universe At Odds with Theory." Accessed June 17, 2018. https://www.scientificamerican.com/article/mature-galaxies-in-young/

Seigel, J. 2014. "America is Getting the Science of Sun Exposure Wrong: Why U.S. Dermatologists Won't Base Recommendations on Skin Color." Accessed April 10, 2019. http://nautil.us/issue/14/mutation/america-is-getting-the-science-of-sun-exposure-wrong

Sibley, A. 2012. "Creationism and Millennialism Among the Church Fathers." *Journal of Creation* 26, no. 3. https://creation.com/creation-millennium-church-fathers

Snelling, A.A. 2008. "Thirty Miles of Dirt in a Day." *Answers Magazine* (Oct. – Dec.): 29 – 30.

—. 2005. "Radiohalos in Granites: Evidence for Accelerated Nuclear Decay." In *Radioisotopes and the Age of the Earth: Results of a Young-Earth Creationist Research Initiative*, edited by L. Vardiman, A.A. Snelling and E.F. Chaffin. El Cajon, California: Institute for Creation Research and Chino Valley, Arizona: Creation Research Society.

Stanley, S.M., and Lucza, J.A. 2015. *Earth System History: Fourth Edition.* New York. W.H. Freeman and Company.

Steen, D. 2011. "Aging Rattle Snakes: Don't Bother Counting Their Rattles." Accessed May 8, 2019. http://www.livingalongsidewildlife.com/2011/07/aging-rattlesnakes-dont-bother-counting.html

Stephenson, F.R. 1981. "Tidal Recession of the Moon from Ancient and Modern Data." *Journal of the British Astronomical Association* 91: 136 – 147.

Stern, R.J., and Scholl, D.W. 2010 "Yin and Yang of Continental Crust Creation and Destruction by Plate Tectonic Processes." *International Geology Review* 52, no. 1: 1 – 31.

Stromberg, J. 2013. "The Sun's Magnetic Field is About to Reverse." Accessed February 22, 2019. https://www.smithsonianmag.com/science-nature/the-suns-magnetic-field-is-about-to-reverse-3738753/

University of California Museum of Paleontology. 1996. "Evolution and Paleontology in the Ancient World. Accessed March 14, 2019. https://ucmp.berkeley.edu/history/ancient.html

Westermann, C. 1994. *Genesis 1 – 11: A Continental Commentary.* Translated by John J. Scullion. Minneapolis. Fortress Press.

Whitmore, J.H., and P.A., Garner. 2018. "The Coconino Sandstone (Permian, Arizona, USA): Implications for the Origin of Ancient Cross-bedded Sandstones." In *Proceedings of the Eighth International Conference on Creationism*, edited by J.H. Whitmore, 581 – 627. Pittsburgh, Pennsylvania: Creation Science Fellowship.

Whitmore, J.H., Forsythe, G., and Garner, P.A. 2015. "Intraformational Parabolic Recumbent Folds in the Coconino Sandstone (Permian) and Two Other Formations in Sedona, Arizona (USA)." Science and Mathematics Faculty Publications, 308. Accessed May 8, 2019. http://digitalcommons.cedarville.edu/science_and_mathematics_publications/308

Whitmore, J.H., and Strom, R. 2010. "Sand-injectites at the Base of the Coconino Sandstone, Grand Canyon, Arizona (USA)." *Sedimentary Geology* 230: 46 – 59.

Wimmer, R. 2011. "Wood Quality: Causes, Methods, Control." The University of Natural Resources and Life Sciences, Vienna. Accessed November 5, 2018. http://www.boku.ac.at/botanik/wood/woodquality/Chapter2.pdf

Wise, K.P. 2002. *Faith, Form and Time: What the Bible Teaches and Science Confirms About Creation and the Age of the Universe.* Nashville, Tennessee. Broadman and Holman Publishers.

Wood, T.C., and Murray, J.M. 2003. *Understanding the pattern of life: origins and organization of the species.* Nashville, Tennessee. Broadman and Holman Publishers.

Young, Wei., Zuyin, Pu., Qiugang, Zong., Weixing, Wan., Zhipeng, Ren., Markus, Fraenze., Eduard, Dubinin., Feng, Tian., Quanqi, Shi., Suiyan, Fu., Minghua, Honga. 2014. "Oxygen Escape from the Earth During Geomagnetic Reversals: Implications to Mass extinction." *Earth and Planetary Science Letters* 394: 94 – 98. https://doi.org/10.1016/j.epsl.2014.03.018

Made in United States
North Haven, CT
10 November 2026

22106124R00115